# A First Course in Combinatorial Optimization

*A First Course in Combinatorial Optimization* is a text for a one-semester introductory graduate-level course for students of operations research, mathematics, and computer science. It is a self-contained treatment of the subject, requiring only some mathematical maturity. Topics include linear and integer programming, polytopes, matroids and matroid optimization, shortest paths, and network flows.

Central to the exposition is the polyhedral viewpoint, which is the key principle underlying the successful integer-programming approach to combinatorial-optimization problems. Another key unifying topic is matroids. The author does not dwell on data structures and implementation details, preferring to focus on the key mathematical ideas that lead to useful models and algorithms. Problems and exercises are included throughout as well as references for further study.

# Cambridge Texts in Applied Mathematics



The aim of this series is to provide a focus for publishing textbooks in applied mathematics at the advanced undergraduate and beginning graduate levels. It is planned that the books will be devoted to covering certain mathematical techniques and theories and to exploring their applications.

Maximum and Minimum Principles
M.J. Sewell

Solitions
P.G. Drazin and R.S. Johnson

The Kinematics of Mixing
J.M. Ottino

Introduction to Numerical Linear Algebra and Optimisation
Phillippe G. Ciarlet

Integral Equations
David Porter and David S.G. Stirling

Perturbation Methods
E.J. Hinch

The Thermomechanics of Plasticity and Fracture
Gerard A. Maugin

Boundary Integral and Singularity Methods for Linearized Viscous Flow
C. Pozrikidis

Nonlinear Systems
P.G. Drazin

Stability, Instability and Chaos
Paul Glendinning

Applied Analysis of the Navier–Stokes Equations
C.R. Doering and J.D. Gibbon

Viscous Flow
H. Ockendon and J.R. Ockendon

Similarity, Self-similarity and Intermediate Asymptotics
G.I. Barenblatt

A First Course in the Numerical Analysis of Differential Equations
A. Iserles

Complex Variables: Introduction and Applications
Mark J. Ablowitz and Athanssios S. Fokas

# A First Course in Combinatorial Optimization

JON LEE

*IBM T.J. Watson Research Center, Yorktown Heights, New York*

CAMBRIDGE
UNIVERSITY PRESS

PUBLISHED BY THE PRESS SYNDICATE OF THE UNIVERSITY OF CAMBRIDGE
The Pitt Building, Trumpington Street, Cambridge, United Kingdom

CAMBRIDGE UNIVERSITY PRESS
The Edinburgh Building, Cambridge CB2 2RU, UK
40 West 20th Street, New York, NY 10011-4211, USA
477 Williamstown Road, Port Melbourne, VIC 3207, Australia
Ruiz de Alarcón 13, 28014 Madrid, Spain
Dock House, The Waterfront, Cape Town 8001, South Africa

http://www.cambridge.org



First published 2004

Printed in the United States of America

*Typeface* Times Roman 10/13 pt. *System* LaTeX $2_\varepsilon$ [TB]

*A catalog record for this book is available from the British Library.*

*Library of Congress Cataloging in Publication data available*

ISBN 0 521 81151 1 hardback
ISBN 0 521 01012 8 paperback

## THE HOUSE JACK BUILT

Open doors so I walk inside
Close my eyes find my place to hide
And I shake as I take it in
Let the show begin

Open my eyes
Just to have them close again
Well on my way
On my way to where I graze
It swallows me
As it takes me in his home
I twist away
As I kill this world

Open doors so I walk inside
Close my eyes find my place to hide
And I shake as I take it in
Let the show begin

Open my eyes
Just to have them close once again
Don't want control
As it takes me down and down again
Is that the moon
Or just a light that lights this deadend street?
Is that you there
Or just another demon that I meet?

The higher you walk
The farther you fall
The longer the walk
The farther you crawl
My body my temple
This temple it tells
"Step into the house that Jack built"

The higher you walk
The farther you fall
The longer the walk
The farther you crawl
My body my temple
This temple it tells
"Yes this is the house that Jack built"

Open doors as I walk inside
Swallow me so the pain subsides
And I shake as I take this in
Let the show begin

The higher you walk
The farther you fall
The longer the walk
The farther you crawl
My body my temple
This temple it tells
"Yes this is the house that Jack built"

The higher you walk
The farther you fall
The longer the walk
The farther you crawl
My body my temple
This temple it tells
"Yes I am I am I am"

Open my eyes
It swallows me
Is that you there
I twist away
Away
Away
Away

– Metallica (Load)

# *Contents*

# *Preface*

This is the house that Jack built. Ralph prepared the lot. There were *many* independent contractors who did beautiful work; some putting on splendid additions. Martin, Laci, and Lex rewired the place. The work continues. But this is the house that Jack built.

This textbook is designed to serve as lecture notes for a one-semester course focusing on combinatorial optimization. I am primarily targeting this at the graduate level, but much of the material may also be suitable for excellent undergraduate students. The goal is to provide an enticing, rigorous introduction to the mathematics of the subject, *within the context of a one-semester course*. There is a strong emphasis on the unifying roles of matroids, submodularity, and polyhedral combinatorics.

I do not pretend that this book is an exhaustive treatment of combinatorial optimization. I do not emphasize data structures, implementation details, or sophisticated approaches that may yield decidedly faster and more practical algorithms. Such are important issues, but I leave them for later independent study. The approach that I take is to focus, mostly, on the beautiful. Also, I note that the terrain of the field shifts rapidly. For example, Gomory's seminal work on integer programming from the 1960s, which was featured prominently in textbooks in the early 1970s, was out of vogue by the late 1970s and through the early 1990s when it was assessed to have no practical value. However, by the late 1990s, Gomory's methods were found to be practically useful. Rather than try and guess as to what will be practically useful some decades from now, I prefer to emphasize some of the work that I regard as foundational.

Also, I do not dwell on applications. To some extent, the applications are the *raison d'être* of combinatorial optimization. However, for the purposes of this book, I take the view that the interesting mathematics and algorithm engineering are justifications enough for studying the subject. Despite (because of?) the fact that I only touch on applications, one can develop talent in modeling and in developing solution methods by working through this book. This apparent paradox is explained by the fact that mathematical abstraction and modeling abstraction are very close cousins.

The prerequisites for studying this book are (1) some mathematical sophistication and (2) *elementary* notions from graph theory (e.g., path, cycle, tree). If one has already studied linear programming, then a good deal of Chapter 0 can be omitted.

*Problems* (requests for short proofs) and *Exercises* (requests for calculations) are interspersed in the book. Each Problem is designed to teach or reinforce a concept. Exercises are used to either verify understanding of an algorithm or to illustrate an idea. Problems and Exercises should be attempted as they are encountered. I have found it to be very valuable to have students or me present correct solutions to the class on each assignment due date. The result is that the text plays longer than the number of pages suggests.

The Appendix should at least be skimmed before working through the main chapters; it consists of a list of notation and terminology that is, for the most part, *not* defined in the main chapters.

A list of references for background and supplementary reading is provided.

Finally, there is a set of indexes that may aid in navigating the book: the first is an index of examples; the second is an index of exercises; the third is an index of problems; the fourth is an index of results (i.e., lemmas, theorems, propositions, corollaries); the last is an index of algorithms.

We begin with an Introduction to the *mind set* of combinatorial optimization and the polyhedral viewpoint.

Chapter 0 contains "prerequisite" results concerning polytopes and linear programming. Although the material of Chapter 0 is prerequisite, most linear-programming courses will not have covered all of this chapter. When I have taught from this book, I start right in with Chapter 1 after working through the Introduction. Then, as needed while working through Chapters 1–8, I ask students to read, or I cover in class, parts of Chapter 0. In particular, Section 0.5 is needed for Sections 1.7, 3.4, and 4.2; Section 0.2 is needed for Sections 1.7, 4.3, 4.4, and 5.2; Section 0.6 is needed for Section 7.3; and Sections 0.3 and 0.7 are needed for Section 6.3.

Although Chapter 0 does not contain a comprehensive treatment of linear programming, by adding some supplementary material on (1) practical implementation details for the simplex method, (2) the ellipsoid method, and (3) interior-point methods, this chapter can be used as the core of a more full treatment of linear programming.

The primary material starts with Chapter 1. In this chapter, we concentrate on matroids and the greedy algorithm. Many of the central ideas that come up later, like submodularity and methods of polyhedral combinatorics, are first explored in this chapter.

Chapter 2, in which we develop the basic algorithms to calculate minimum-weight dipaths, is somewhat of a digression. However, minimum-weight dipaths and the associated algorithms are important building blocks for other algorithms.

In Chapter 3, we discuss the problem of finding maximum-cardinality, and, more generally, maximum-weight sets that are independent in two matroids on a common ground set. The algorithms and polyhedral results are striking in their beauty and complexity.

The subject of Chapter 4 is matchings in graphs. As in the previous chapter, striking algorithms and polyhedral results are presented. We discuss some applications of matching to other combinatorial-optimization problems.

The subjects of Chapters 3 and 4 can be viewed as two different generalizations of the problem of finding maximum-cardinality and maximum-weight matchings in *bipartite* graphs. We find that König's min/max theorem, as well as the algorithmic and polyhedral results, generalize in quite different ways.

In Chapter 5, we discuss the maximum-flow problem for digraphs and related cut problems. Although the topic seems less intricate than those of the two previous chapters, we discuss the seminal method of Edmonds and Karp that is used to produce an efficient algorithm. Also, the methods of this chapter relate to those of Chapter 2.

In Chapter 6, we study cutting-plane methods for integer programming. We begin with the fundamental idea of taking nonnegative linear combinations and rounding. The details of Gomory's finite cutting-plane algorithm are described. There is a general discussion of methods for tightening integer-programming formulations. Examples of special-purpose cutting-plane methods for combinatorial-optimization problems are also given.

In Chapter 7, Branch-&-Bound methods for solving discrete-optimization problems are described. The general framework is not very interesting from a mathematical point of view, but the bounding methods, for example, can be quite sophisticated. Also, Branch-&-Bound is a very useful practical technique for solving combinatorial-optimization problems.

In Chapter 8, we discuss optimization of submodular functions. Many of the problems that were treated in the earlier chapters can be viewed as problems of minimizing or maximizing particular submodular functions. Although the efficient algorithms for minimizing general submodular functions are not described, it is valuable to explore the unifying role of submodular functions.

And there it ends. A sequel to this book would study (1) semidefinite programming formulations of combinatorial-optimization problems and associated interior-point algorithms for the solution of the relaxations, (2) efficient approximation algorithms with performance guarantees for combinatorial-optimization problems, (3) algebraic methods for integer programming, (4) and much more on submodular optimization. The practical significance of these subjects has yet to be firmly established, but the theory is great!

I thank those who first taught me about combinatorics and optimization at Cornell: Lou Billera, Bob Bland, Jack Edmonds, George Nemhauser, Mike Todd, and Les Trotter. Further thanks are due to Carl Lee, François Margot, and students at the University of Kentucky and New York University who worked through drafts of this book; they made many valuable suggestions, most of which I stubbornly ignored.

Finally, this project would never have been completed without the firm yet compassionate guidance of Lauren Cowles, Caitlin Doggart, Katie Hew, and Lara Zoble of Cambridge University Press and Michie Shaw of TechBooks.

JON LEE

Yorktown Heights, New York

# Introduction

A *discrete-optimization problem* is a problem of maximizing a real-valued *objective function* $c$ on a finite set of *feasible solutions* $\mathcal{S}$. Often the set $\mathcal{S}$ naturally arises as a subset of $2^E$ (the set of all subsets of $E$), for some finite *ground set* $E$, in which case we have a *combinatorial-optimization problem*. Of course, there is no *problem* because we can just enumerate all feasible solutions – but we seek to do better. Usually, the feasible solutions are *described* in some concise manner, rather than being explicitly listed. The challenge is to develop algorithms that are provably or practically better than enumerating all feasible solutions.

Applications of discrete-optimization problems arise in industry (e.g., manufacturing and distribution, telecommunication-network design and routing, airline crew scheduling) and in applied sciences (e.g., statistics, physics, and chemistry).

Besides the applications, discrete optimization has aspects that connect it with other areas of mathematics (e.g., algebra, analysis and continuous optimization, geometry, logic, numerical analysis, topology, and, of course, other subdisciplines of discrete mathematics such as graph theory, matroid theory, and enumerative combinatorics) as well as computer science. Thus research in discrete optimization is driven by mathematics as well as by applications.

It is almost always the case that the set of feasible solutions $\mathcal{S}$ is delivered to us *descriptively* rather than by an explicit list. For example, $\mathcal{S}$ might be the set of spanning trees of a connected graph. As a complete graph on $n$ vertices has $n^{n-2}$ spanning trees (a nontrivial fact discovered by Cayley), it may come as quite a surprise that finding a 'maximum-weight' spanning tree is about as difficult as sorting the $\binom{n}{2} = n(n-1)/2$ edge weights. As another example, $\mathcal{S}$ might be the set of 'traveling-salesperson's tours' through $n$ points in some metric space. There are $(n-1)!/2$ (equivalence classes of) such tours (as we may call any of the $n$ points the initial point of the tour, and we can reverse the ordering of the points to obtain another tour of the same total length). The problem of finding a shortest traveling-salesperson's tour is a notoriously difficult problem; yet we will touch on techniques that enable us to find good solutions for instances that are significantly larger than brute-force enumeration would permit.

An algorithm is *theoretically efficient* for a class of problems if the number of computational steps required for solving instances of the problem is bounded by a polynomial in the number of bits required for encoding the problem (in which integers are encoded in base 2). We encode a rational number by encoding its integer numerator and denominator. This model does not permit the encoding of irrational numbers. To make all of this precise, we would need to carefully specify a model of computation (e.g., the Turing machine). Then, through notions of problem equivalence (e.g., polynomial-time reductions), we would define complexity classes (e.g., the class NP) and the idea of "completeness" for a complexity class. We will hardly touch on such issues in what follows, but a full appreciation of combinatorial optimization, from the point of view of "theoretical efficiency," requires such ideas.

The beacon of theoretical efficiency has its faults as an indicator of practical performance: (1) It is an asymptotic theory, (2) it is a worst-case theory, and (3) the order of the bounding polynomial may be quite high. Correspondingly, we note that (1) practical problems have some limited size, (2) practical instances may be quite different than worst-case instances, and (3) a high-order polynomial may grow too quickly in the limited range of problem sizes that are of practical concern. Still, this guiding light has shown the way to many practical methods.

For combinatorial-optimization problems, it will often be enlightening, and sometimes computationally effective, to embed our problem in $\mathbf{R}^E$ (real $|E|$-dimensional space with coordinates indexed by $E$). The natural method is as follows. We consider the convex hull $\mathcal{P}_\mathcal{S}$ of the set of characteristic vectors of sets in $\mathcal{S}$ – that is, the smallest convex set that contains these characteristic vectors. Next, we need to find a function $\tilde{c} : [0, 1]^E \mapsto \mathbf{R}$ such that, if $x(S)$ is the characteristic vector of a feasible set $S$, then $\tilde{c}(x(S)) = c(S)$. The success of such an approach depends, critically, on the form of the objective function. Concave functions are relatively easy to maximize (provided we have a description of $\mathcal{P}_\mathcal{S}$ as the solution set of linear inequalities), as in this case a local maximum is a global maximum. On the other hand, convex functions have the nice property that they are maximized by extreme points of a polytope – these extreme points are characteristic vectors of our feasible sets. For linear functions we have the best of both worlds. A *weight function* $c : 2^E \mapsto \mathbf{R}$ satisfies $c(S) = \sum_{e \in S} c(e)$, $\forall\ S \subset E$ [we take the liberty of writing $c(e)$ for $c(\{e\})$]. The weight function $c$ naturally leads to the linear function $\tilde{c}$ defined by $\tilde{c}(x) = \sum_{e \in E} c(e) x_e$, $\forall\ x \in \mathbf{R}^E$; note that $c(S) = \tilde{c}(x(S))$. Most of the combinatorial-optimization problems that we will study involve optimizing weight functions. This does not mean that we can easily solve all combinatorial-optimization problems involving the optimization of weight functions. The challenge in the approach that has been outlined is to find a useful description of $\mathcal{P}_{\mathcal{S}}$ by means of linear inequalities.

Next, we look at a concrete example. To visualize the geometry of the example, we are forced to use an instance with very few elements in the ground set. Our ground set $E := \{1, 2, 3\}$ corresponds to the set of edges of the following graph:

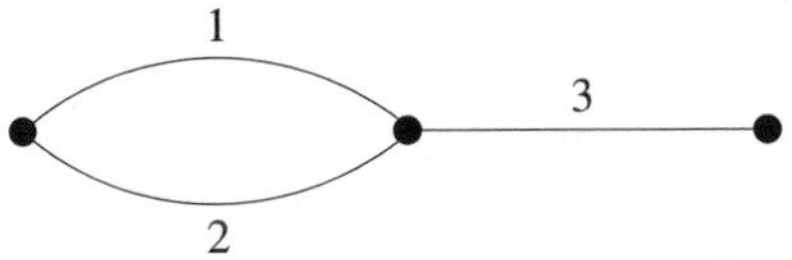


We define our set $\mathcal{S}$ of feasible sets to consist of subsets of $E$ that are acyclic (i.e., contain no cycle). That is, $\mathcal{S}$ is the set of *forests* of the graph. Here

$$\mathcal{S} = \{\emptyset, \{1\}, \{2\}, \{3\}, \{1, 3\}, \{2, 3\}\}$$

(the only sets containing cycles are $\{1, 2\}$ and $\{1, 2, 3\}$).

We consider the characteristic vectors of sets in $\mathcal{S}$, namely,

$$(0, 0, 0),$$
$$(1, 0, 0),$$
$$(0, 1, 0),$$
$$(0, 0, 1),$$
$$(1, 0, 1),$$
$$(0, 1, 1).$$

Next, we embed these points in $\mathbf{R}^E$, and we depict the convex hull $\mathcal{P}_{\mathcal{S}}$:

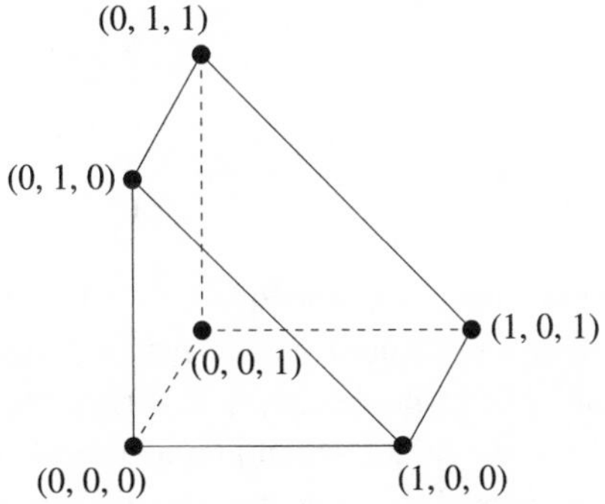

The polytope $\mathcal{P}_S$ is one-half of the unit cube. It is the solution set of the linear inequalities

$$
\begin{aligned}
x_1 &\geq 0, \\
x_2 &\geq 0, \\
1 \geq x_3 &\geq 0, \\
x_1 + x_2 &\leq 1 .
\end{aligned}
$$

If, for example, we maximize the linear function $5x_1 + 4x_2 + x_3$ over the solutions to this inequality system, we get the optimal solution $x = (1, 0, 1)$, which is the characteristic vector of $\{1, 3\}$.

We may not be so fortunate if we model the points carelessly. For example, the set of linear inequalities

$$
\begin{aligned}
1 \geq x_1 &\geq 0, \\
1 \geq x_2 &\geq 0, \\
1 \geq x_3 &\geq 0, \\
x_1 + x_2 - x_3 &\leq 1, \\
x_1 + x_2 + x_3 &\leq 2,
\end{aligned}
$$

has precisely the same $0/1$ solutions as the inequality system that describes $\mathcal{P}_S$. It is easy to see that $(1, 1, 0)$ (the characteristic vector of $\{1, 2\}$) is the only $0/1$ vector excluded by $x_1 + x_2 - x_3 \leq 1$. Also, $(1, 1, 1)$ (the characteristic vector of $\{1, 2, 3\}$) is the only $0/1$ vector excluded by $x_1 + x_2 + x_3 \leq 2$. However, these inequalities describe a region that properly contains $\mathcal{P}_S$:

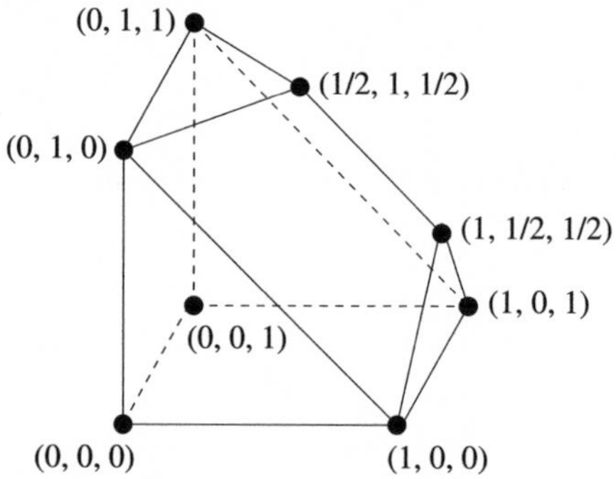


The difficulty with this latter set of inequalities is that there are linear functions having their unique maximum on the feasible region at a point with fractional components. For example, $5x_1 + 4x_2 + x_3$ (the objective function that we used earlier) has its unique maximum at $x = (1, 1/2, 1/2)$. So, if we do not work with the inequalities that describe the convex hull, we cannot rely on linear programming to identify an optimal solution to the underlying combinatorial-optimization problem. Finally, we note that if we add $1/2$ of each of the inequalities

$$x_1 + x_2 - x_3 \leq 1,$$
$$x_1 + x_2 + x_3 \leq 2,$$

we get

$$x_1 + x_2 \leq 3/2.$$

Rounding down the right-hand-side constant, which we can do because the left-hand side will only take on integer values on $0/1$ solutions, we recover the inequality

$$x_1 + x_2 \leq 1,$$

which is needed in the inequality description of $\mathcal{P}_{\mathcal{S}}$.

Even if we have a description of the convex hull using linear inequalities, the situation is far more difficult for nonlinear maximization. For example, it is not hard to see that the function $2x_1 + x_2 - 3x_1x_2 + x_3$ is maximized on $\mathcal{P}_{\mathcal{S}}$ by the point $x = (1, 0, 1)$. However, this function is not concave, so methods of nonlinear optimization that would seek a local minimum on the convex set $\mathcal{P}_{\mathcal{S}}$ may fail to find the optimum. For example, the point $x = (0, 1, 1)$ is a strict local minimizer on $\mathcal{P}_{\mathcal{S}}$. Therefore, it is hard to proceed from that point to the optimum by use of local methods of nonlinear programming. We can try to salvage something by transforming the objective function. The concave function $-3x_1^2 - 3x_2^2 - 3x_1x_2 + 5x_1 + 4x_2 + x_3$ takes on the same values at 0/1 vectors as the original function (we are just using the identity $x_j^2 = x_j$ when $x_j$ is 0 or 1). This function has its unique maximum on $\mathcal{P}_{\mathcal{S}}$ at $x = (2/3, 1/3, 1)$. However, this point is not a characteristic vector. Therefore, even though it is relatively easy to find this maximum by continuous local-search methods of nonlinear programming (maximizing a strictly concave function on a concave set is a situation in which finding a local maximizer is sufficient), the solution does not solve the underlying combinatorial-optimization problem. Finally, if we are clever enough to notice that the function $2x_1 + x_2 + x_3$ takes on the same values at *feasible* 0/1 vectors as the original function $2x_1 + x_2 - 3x_1x_2 + x_3$, then we can easily find $x = (1, 0, 1)$ as the solution of a linear program.

The important point to draw from this example is that continuous modeling must be done very carefully when variables are used to represent discrete choices in a combinatorial-optimization problem. This section closes with some Exercise and Problems that further develop this point.

**Exercise (Comparing relaxations).** The following three systems of inequalities have the same set of *integer-valued* solutions.

$$(I)\quad \begin{cases} x_1 + x_2 \le 1 & x_1 \ge 0 \\ x_1 + x_3 \le 1 & x_2 \ge 0 \\ x_1 + x_4 \le 1 & x_3 \ge 0 \\ x_2 + x_3 \le 1 & x_4 \ge 0 \\ x_2 + x_4 \le 1 & \end{cases}$$

$$(II)\quad \begin{cases} 2x_1 + 2x_2 + x_3 + x_4 \le 2 \\ 0 \le x_1 \le 1 \\ 0 \le x_2 \le 1 \\ 0 \le x_3 \le 1 \\ 0 \le x_4 \le 1 \end{cases} \qquad (III)\quad \begin{cases} x_1 + x_2 + x_3 \le 1 \\ x_1 + x_2 + x_4 \le 1 \\ x_1 \ge 0 \\ x_2 \ge 0 \\ x_3 \ge 0 \\ x_4 \ge 0 \end{cases}$$

In fact, the solutions to each system are the characteristic vectors of the "vertex packings" of the graph following – a *vertex packing* of $G$ is just a set of vertices $S$ with no edges of $G$ between elements of $S$:

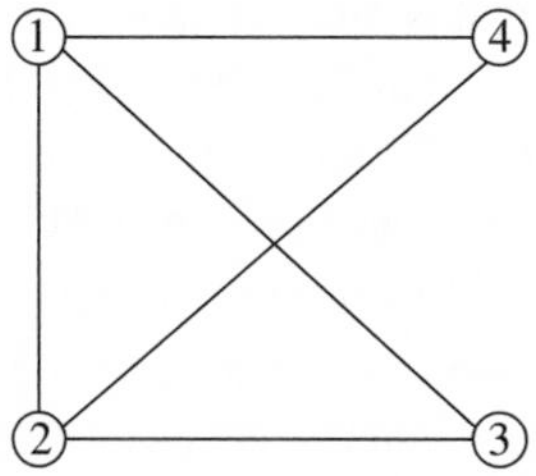


Compare how closely the three systems of inequalities approximate the set of integer-valued solutions in real space.

**Problem (Uncapacitated facility location).** The *uncapacitated facility-location problem* involves production and distribution of a single commodity at available facility locations, numbered $1, 2, \ldots, n$. Customers, numbered $1, 2, \ldots, m$ have demand for the commodity. A fixed-cost $f_i$ is incurred if any (positive) production occurs at facility $i$. The profit (not accounting for the fixed costs) for satisfying the fraction $x_{ij}$ of the demand of customer $j$ from facility $i$ is $c_{ij}x_{ij}$. The goal is to maximize net profit, subject to satisfying all customer demand exactly. We can formulate this problem as the program

$$\max -\sum_{i=1}^{m} f_i y_i + \sum_{i=1}^{m}\sum_{j=1}^{n} c_{ij} x_{ij}$$

subject to:

$$\sum_{i=1}^{m} x_{ij} = 1, \text{ for } j = 1, 2, \ldots, n;$$

$$(*) \qquad -ny_i + \sum_{j=1}^{n} x_{ij} \le 0, \text{ for } i = 1, 2, \ldots, m;$$

$$0 \le x_{ij} \le 1, \text{ for } i = 1, 2, \ldots, m \text{ and } j = 1, 2, \ldots, n;$$

$$0 \le y_i \le 1 \text{ integer, for } i = 1, 2, \ldots, m.$$

Compare the strength of $(*)$ and

$$(**) \qquad -y_i + x_{ij} \le 0, \text{ for } i = 1, 2, \ldots, m \text{ and } j = 1, 2, \ldots, n.$$

**Problem (Piecewise-linear functions).** In practical instances of many optimization problems, key quantities, like costs, may not be well modeled as linear functions. In many instances, however, a piecewise-linear function is adequate. Let $x^1 < x^2 < \cdots < x^n$ be real numbers. We consider the piecewise-linear function $f : [x^1, x^n] \mapsto \mathbf{R}$ that we define by linearly interpolating $f$ between the $x^i$. That is, if $x = \lambda_i x^i + \lambda_{i+1} x^{i+1}$, for some $\lambda_i, \lambda_{i+1} \ge 0$ with $\lambda_i + \lambda_{i+1} = 1$, then $f(x) := \lambda_i f(x^i) + \lambda_{i+1} f(x^{i+1})$:

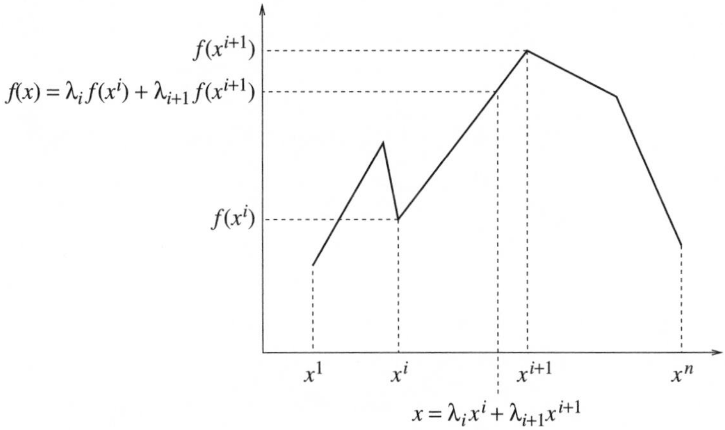

The difficulty in formulating this with linear constraints is that the choice of $i$ depends on $x$. Still, with 0/1 variables, we can make the choice. We employ the formulation

$$f(x) = \sum_{i=1}^{n} \lambda_i f(x^i);$$

$$\sum_{i=1}^{n} \lambda_i = 1; \qquad \sum_{i=1}^{n-1} y_i = 1;$$

$$\lambda_i \geq 0, \text{ for } i = 1, 2, \ldots, n; \quad y_i \geq 0 \text{ integer, for } i = 1, 2, \ldots, n-1;$$

$$(*) \quad y_i = 1 \Longrightarrow \text{only } \lambda_i \text{ and } \lambda_{i+1} \text{ may be positive.}$$

a. Explain why (∗) can be modeled by

$$(**) \quad \begin{cases} \lambda_1 \leq y_1 \\ \lambda_i \leq y_{i-1} + y_i, \text{ for } i = 2, 3, \ldots, n-1. \\ \lambda_n \leq y_{n-1} \end{cases}$$

b. Compare the strength of (∗∗) and

$$(***) \quad \begin{cases} \sum_{i=1}^{j} y_i \leq \sum_{i=1}^{j+1} \lambda_i, \text{ for } j = 1, 2, \ldots, n-2 \\ \sum_{i=j}^{n-1} y_i \leq \sum_{i=j}^{n} \lambda_i, \text{ for } j = 2, 3, \ldots, n-1 \end{cases}.$$

# 0

# Polytopes and Linear Programming

In this chapter, we review many of the main ideas and results concerning polytopes and linear programming. These ideas and results are prerequisite to much of combinatorial optimization.

## 0.1 Finite Systems of Linear Inequalities

Let $x^k$, $k \in N$, be a finite set of points in $\mathbf{R}^n$. Any point $x \in \mathbf{R}^n$ of the form $x = \sum_{k \in N} \lambda_k x^k$, with $\lambda_k \in \mathbf{R}$, is a *linear combination* of the $x^k$. If, in addition, we have all $\lambda_k \geq 0$, then the combination is *conical*. If $\sum_{k \in N} \lambda_k = 1$, then the combination is *affine*. Combinations that are both conical and affine are *convex*. The points $x^k \in \mathbf{R}^n$, $k \in N$, are *linearly independent* if $\sum_{k \in N} \lambda_k x^k = 0$ implies $\lambda_k = 0 \; \forall \, k \in N$. The points $x^k \in \mathbf{R}^n$, $k \in N$, are *affinely independent* if $\sum_{k \in N} \lambda_k x^k = 0$, $\sum_{k \in N} \lambda_k = 0$ implies $\lambda_k = 0 \; \forall \, k \in N$. Equivalently, the points $x^k \in \mathbf{R}^n$, $k \in N$, are *affinely independent* if the points $\binom{x^k}{1} \in \mathbf{R}^{n+1}$, $k \in N$, are *linearly independent*.

A set $X \subset \mathbf{R}^n$ is a *subspace/cone/affine set/convex set* if it is closed under (finite) linear/conical/affine/convex combinations. The *linear span/conical hull/affine span/convex hull* of $X$, denoted $\mathrm{sp}(X)/\mathrm{cone}(X)/\mathrm{aff}(X)/\mathrm{conv}(X)$, is the set of all (finite) linear/conical/affine/convex combinations of points in $X$. Equivalently, and this needs a proof, $\mathrm{sp}(X)/\mathrm{cone}(X)/\mathrm{aff}(X)/\mathrm{conv}(X)$ is the intersection of all subspaces/cones/affine sets/convex sets containing $X$.

In the following small example, $X$ consists of two linearly independent points:

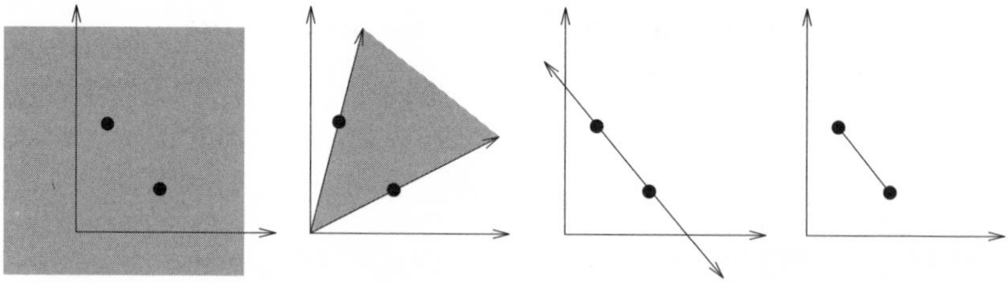

sp($X$) cone($X$) aff($X$) conv($X$)

A *polytope* is conv($X$) for some finite set $X \subset \mathbf{R}^n$. A polytope can also be described as the solution set of a finite system of linear inequalities. This result is called Weyl's Theorem (for polytopes). A precise statement of the theorem is made and its proof is provided, after a very useful elimination method is described for linear inequalities.

Fourier–Motzkin Elimination is the analog of Gauss–Jordan Elimination, but for linear *in*equalities. Consider the linear-inequality system

$$\sum_{j=1}^{n} a_{ij} x_j \le b_i \ , \ \text{for } i = 1, 2, \ldots, m.$$

Select a variable $x_k$ for elimination. Partition $\{1, 2, \ldots, m\}$ based on the signs of the numbers $a_{ik}$, namely,

$$\begin{aligned} S_+ &:= \{i \ : \ a_{ik} > 0\}, \\ S_- &:= \{i \ : \ a_{ik} < 0\}, \\ S_0 &:= \{i \ : \ a_{ik} = 0\}. \end{aligned}$$

The new inequality system consists of

$$\sum_{j=1}^{n} a_{ij} x_j \le b_i \text{ for } i \in S_0,$$

together with the inequalities

$$-a_{lj} \left( \sum_{j=1}^{n} a_{ij} x_j \le b_i \right) + a_{ij} \left( \sum_{j=1}^{n} a_{lj} x_j \le b_l \right),$$

for all pairs of $i \in S_+$ and $l \in S_-$. It is easy to check that

1. the new system of linear inequalities does not involve $x_k$;
2. each inequality of the new system is a nonnegative linear combination of the inequalities of the original system;

3. if

$$(x_1, x_2, \ldots, x_{k-1}, x_k, x_{k+1}, \ldots, x_n)$$

solves the original system, then

$$(x_1, x_2, \ldots, x_{k-1}, x_{k+1}, \ldots, x_n)$$

solves the new system;

4. if

$$(x_1, x_2, \ldots, x_{k-1}, x_{k+1}, \ldots, x_n)$$

solves the new system, then

$$(x_1, x_2, \ldots, x_{k-1}, x_k, x_{k+1}, \ldots, x_n)$$

solves the original system, for some choice of $x_k$.

Geometrically, the solution set of the new system is the projection of the solution set of the original system, in the direction of the $x_k$ axis.

Note that if $S_+ \cup S_0$ or $S_- \cup S_0$ is empty, then the new inequality system has no inequalities. Such a system, which we think of as still having the remaining variables, is solved by every choice of $x_1, x_2, \ldots, x_{k-1}, x_{k+1}, \ldots, x_n$.

**Weyl's Theorem (for polytopes).** *If $\mathcal{P}$ is a polytope then*

$$\mathcal{P} = \left\{ x \in \mathbf{R}^n \ : \ \sum_{j=1}^{n} a_{ij}x_j \le b_i, \ \ \textit{for } i = 1, 2, \ldots, m \right\},$$

*for some positive integer $m$ and choice of real numbers $a_{ij}$, $b_i$, $1 \le i \le m$, $1 \le j \le n$.*

*Proof.* Weyl's Theorem is easily proved by use of Fourier–Motzkin Elimination. We consider the linear-inequality system obtained from

$$x_j - \sum_{k \in N} \lambda_k x_j^k = 0, \text{ for } j = 1, 2, \ldots, n;$$
$$\sum_{k \in N} \lambda_k = 1;$$
$$\lambda_k \ge 0, \quad \forall \quad k \in N,$$

by replacing each equation with a pair of inequalities. Note that, in this system, the numbers $x_j^k$ are constants. We apply Fourier–Motzkin Elimination to the linear-inequality system, so as to eliminate all of the $\lambda_k$ variables. The final system must include all of the variables $x_j$, $j \in N$, because otherwise $\mathcal{P}$ would be unbounded. Therefore, we are left with a nonempty linear-inequality system in the $x_j$, $j \in N$, describing $\mathcal{P}$. ∎

Also, it is a straightforward exercise, by use of Fourier–Motzkin Elimination, to establish the following theorem characterizing when a system of linear inequalities has a solution.

**Theorem of the Alternative for Linear Inequalities.** *The system*

$$(I) \qquad \sum_{j=1}^{n} a_{ij}x_j \le b_i, \textit{ for } i = 1, 2, \ldots, m$$

*has a solution if and only if the system*

$$(II) \qquad \begin{aligned} \sum_{i=1}^{m} y_i a_{ij} &= 0, \textit{ for } j = 1, 2, \ldots, n; \\ y_i &\ge 0, \textit{ for } i = 1, 2, \ldots, m; \\ \sum_{i=1}^{m} y_i b_i &< 0 \end{aligned}$$

*has no solution.*

*Proof.* Clearly we cannot have solutions to both systems because that would imply

$$0 = \sum_{j=1}^{n} x_j \sum_{i=1}^{m} y_i a_{ij} = \sum_{i=1}^{m} y_i \sum_{j=1}^{n} a_{ij}x_j \le \sum_{i=1}^{m} y_i b_i < 0.$$

Now, suppose that $I$ has no solution. Then, after eliminating all $n$ variables, we are left with an inconsistent system in no variables. That is,

$$\sum_{j=1}^{n} 0 \cdot x_j \le b'_i, \text{ for } i = 1, 2, \ldots, p,$$

where, $b'_k < 0$ for some $k$, $1 \le k \le p$. The inequality

$$(*) \qquad \sum_{j=1}^{n} 0 \cdot x_j \le b'_k,$$

is a nonnegative linear combination of the inequality system $I$. That is, there exist $y_i \ge 0$, $i = 1, 2, \ldots, m$, so that $(*)$ is just

$$\sum_{i=1}^{m} y_i \left( \sum_{j=1}^{n} a_{ij}x_j \le b_i \right).$$

Rewriting this last system as

$$\sum_{j=1}^{n} \left( \sum_{i=1}^{m} y_i a_{ij} \right) x_j \le \sum_{i=1}^{m} y_i b_i,$$

and equating coefficients with the inequality $(*)$, we see that $y$ is a solution to $(*)$. ■

An equivalent result is the "Farkas Lemma."

**Theorem (Farkas Lemma).** *The system*

$$\sum_{j=1}^{n} a_{ij}\, x_j = b_i, \textit{ for } i = 1, 2, \ldots, m;$$
$$x_j \geq 0, \textit{ for } j = 1, 2, \ldots, n.$$

*has a solution if and only if the system*

$$\sum_{i=1}^{m} y_i a_{ij} \geq 0, \textit{ for } j = 1, 2, \ldots, n;$$
$$\sum_{i=1}^{m} y_i b_i < 0$$

*has no solution.*

The Farkas Lemma has the following geometric interpretation: Either $b$ is in the conical hull of the columns of $A$, or there is a vector $y$ that makes a nonobtuse angle with every column of $A$ and an obtuse angle with $b$ (but not both):

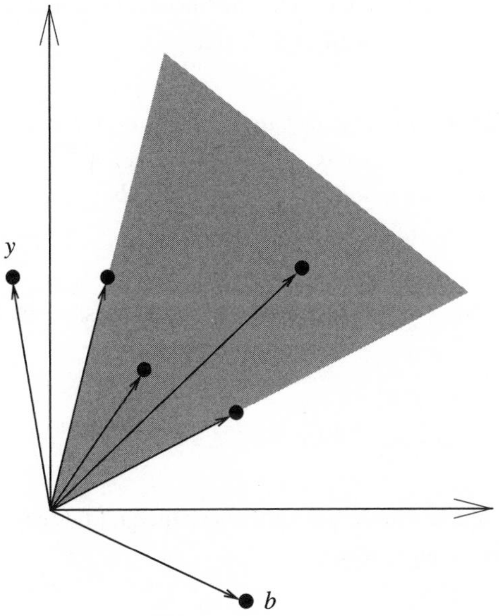

**Problem (Farkas Lemma).** Prove the Farkas Lemma by using the Theorem of the Alternative for Linear Inequalities.

## 0.2 Linear-Programming Duality

Let $c_j$, $1 \le j \le n$, be real numbers. Using real variables $x_j$, $1 \le j \le n$, and $y_i$, $1 \le i \le m$, we formulate the dual pair of linear programs:

$$(P) \qquad \begin{aligned} &\max \sum_{j=1}^{n} c_j x_j \\ &\text{subject to:} \\ &\qquad \sum_{j=1}^{n} a_{ij} x_j \le b_i, \quad \text{for } i = 1, 2, \ldots, m, \end{aligned}$$

$$(D) \qquad \begin{aligned} &\min \sum_{i=1}^{m} y_i b_i \\ &\text{subject to:} \\ &\qquad \sum_{i=1}^{m} y_i a_{ij} = c_j, \quad \text{for } i = 1, 2, \ldots, n; \\ &\qquad\qquad y_i \ge 0, \quad \text{for } i = 1, 2, \ldots, m. \end{aligned}$$

As the following results make clear, the linear programs $P$ and $D$ are closely related in their solutions as well as their data.

**Weak Duality Theorem.** *If $x$ is feasible to $P$ and $y$ is feasible to $D$, then $\sum_{j=1}^{n} c_j x_j \le \sum_{i=1}^{m} y_i b_i$. Hence,* (1) *if $\sum_{j=1}^{n} c_j x_j = \sum_{i=1}^{m} y_i b_i$, then $x$ and $y$ are optimal, and* (2) *if either program is unbounded, then the other is infeasible.*

*Proof.* Suppose that $x$ is feasible to $P$ and $y$ is feasible to $D$. Then we see that

$$\sum_{j=1}^{n} c_j x_j = \sum_{j=1}^{n} \left( \sum_{i=1}^{m} y_i a_{ij} \right) x_j = \sum_{i=1}^{m} y_i \left( \sum_{j=1}^{n} a_{ij} x_j \right) \le \sum_{i=1}^{m} y_i b_i. \qquad \blacksquare$$

The Theorem of the Alternative for Linear Inequalities can be used to establish the following stronger result.

**Strong Duality Theorem.** *If P and D have feasible solutions, then they have optimal solutions* $x, y$ *with* $\sum_{j=1}^{n} c_j x_j = \sum_{i=1}^{m} y_i b_i$. *If either is infeasible, then the other is either infeasible or unbounded.*

*Proof.* We consider the system of linear inequalities:

$$
(I)\qquad
\begin{array}{rll}
\sum_{j=1}^{n} a_{ij} x_j & \le b_i, & \text{for } i = 1, 2, \ldots, m; \\
\sum_{i=1}^{m} y_i a_{ij} & \le c_j, & \text{for } j = 1, 2, \ldots, n; \\
-\sum_{i=1}^{m} y_i a_{ij} & \le -c_j, & \text{for } j = 1, 2, \ldots, n; \\
-y_i & \le 0, & \text{for } i = 1, 2, \ldots, m; \\
-\sum_{j=1}^{n} c_j x_j + \sum_{i=1}^{m} b_i y_i & \le 0. &
\end{array}
$$

By the Weak Duality Theorem, it is easy to see that $x \in \mathbf{R}^n$ and $y \in \mathbf{R}^m$ are optimal to $P$ and $D$, respectively, if and only if $(x, y)$ satisfies $I$. By the Theorem of the Alternative for Linear Inequalities, system $I$ has a solution if and only if the system

$$
(II)\qquad
\begin{array}{rll}
\sum_{i=1}^{m} u_i a_{ij} - c_j \tau & = 0, & \text{for } j = 1, 2, \ldots, n; \\
\sum_{j=1}^{n} a_{ij} v_j - \sum_{j=1}^{n} a_{ij} v'_j - s_i + b_i \tau & = 0, & \text{for } i = 1, 2, \ldots, m; \\
u_i & \ge 0, & \text{for } i = 1, 2, \ldots, m; \\
v_j & \ge 0, & \text{for } j = 1, 2, \ldots, n; \\
v'_j & \ge 0, & \text{for } j = 1, 2, \ldots, n; \\
s_i & \ge 0, & \text{for } i = 1, 2, \ldots, m; \\
\tau & \ge 0; & \\
\sum_{i=1}^{m} b_i u_i + \sum_{j=1}^{n} c_j v_j - \sum_{j=1}^{n} c_j v'_j & < 0 &
\end{array}
$$

has no solution. Making the substitution, $h_j := v'_j - v_j$, for $j = 1, 2, \ldots, n$, we get the equivalent system:

$$
(II')\qquad
\begin{aligned}
&\tau \ge 0;\\
&\sum_{i=1}^{m} u_i a_{ij} = c_j \tau, \quad \text{for } j = 1, 2, \ldots, n;\\
&u_i \ge 0, \quad \text{for } i = 1, 2, \ldots, m;\\
&\sum_{j=1}^{n} a_{ij} h_j \le b_i \tau, \quad \text{for } i = 1, 2, \ldots, m;\\
&h_j \ge 0, \quad \text{for } j = 1, 2, \ldots, n;\\
&\sum_{i=1}^{m} b_i u_i < \sum_{j=1}^{n} c_j h_j
\end{aligned}
$$

First, we suppose that $P$ and $D$ are feasible. The conclusion that we seek is that $I$ is feasible. If not, then $II'$ has a feasible solution. We investigate two cases:

*Case 1:* $\tau > 0$ in the solution of $II'$. Then we consider the points $x \in \mathbf{R}^n$ and $y \in \mathbf{R}^m$ defined by $x_j := \frac{1}{\tau} h_j$, for $j = 1, 2, \ldots, n$, and $y_i := \frac{1}{\tau} u_i$, for $i = 1, 2, \ldots, m$. In this case, $x$ and $y$ are feasible to $P$ and $D$, respectively, but they violate the conclusion of the Weak Duality Theorem.

*Case 2:* $\tau = 0$ in the solution to $II'$. Then we consider two subcases.

*Subcase a:* $\sum_{i=1}^{m} u_i b_i < 0$. Then we take any $y \in \mathbf{R}^m$ that is feasible to $D$, and we consider $y' \in \mathbf{R}^m$, defined by $y'_i := y_i + \lambda u_i$, for $i = 1, 2, \ldots, m$. It is easy to check that $y'$ is feasible to $D$, for all $\lambda \ge 0$, and that the objective function of $D$, evaluated on $y'$, decreases without bound as $\lambda$ increases.

*Subcase b:* $\sum_{i=1}^{m} c_j h_j > 0$. Then we take any $x \in \mathbf{R}^n$ that is feasible to $P$, and we consider $x' \in \mathbf{R}^n$, defined by $x'_j := x_j + \lambda h_j$, for $j = 1, 2, \ldots, n$. It is easy to check that $x'$ is feasible to $P$, for all $\lambda \ge 0$, and that the objective function of $P$, evaluated on $x'$, increases without bound as $\lambda$ increases.

In either subcase, we contradict the Weak Duality Theorem.

In either Case 1 or 2, we get a contradiction. Therefore, if $P$ and $D$ are feasible, then $I$ must have a solution – which consists of optimal solutions to $P$ and $D$ having the same objective value.

Next, we suppose that $P$ is infeasible. We seek to demonstrate that $D$ is either infeasible or unbounded. Toward this end, we suppose that $D$ is feasible, and we seek to demonstrate that then it must be unbounded. By the Theorem of the

Alternative for Linear Inequalities, the infeasibility of $P$ implies that the system

$$\sum_{i=1}^{m} u_i a_{ij} = 0, \text{ for } j = 1, 2, \ldots, n;$$
$$u_i \geq 0, \text{ for } i = 1, 2, \ldots, m;$$
$$\sum_{i=1}^{m} u_i b_i < 0$$

has a solution. Taking such a solution $u \in \mathbf{R}^m$ and a feasible solution $y \in \mathbf{R}^m$ to $D$, we proceed exactly according to the recipe in the preceding Subcase a to demonstrate that $D$ is unbounded.

Next, we suppose that $D$ is infeasible. We seek to demonstrate that $P$ is either infeasible or unbounded. Toward this end, we suppose that $P$ is feasible, and we seek to demonstrate that then it must be unbounded. By the Theorem of the Alternative for Linear Inequalities, the infeasibility of $P$ implies that the system

$$\sum_{j=1}^{m} a_{ij} h_j \leq 0, \text{ for } i = 1, 2, \ldots, m;$$
$$\sum_{j=1}^{n} c_j h_j > 0$$

has a solution. Taking such a solution $h \in \mathbf{R}^n$ and a feasible solution $x \in \mathbf{R}^n$ to $P$, we proceed exactly according to the recipe in the preceding Subcase b to demonstrate that $P$ is unbounded. ■

**Problem (Theorem of the Alternative for Linear Inequalities).** Prove the Theorem of the Alternative for Linear Inequalities from the Strong Duality Theorem. *Hint:* Consider the linear program

$$\min \sum_{i=1}^{m} y_i b_i$$

$(D_0)$ subject to:

$$\sum_{i=1}^{m} y_i a_{ij} = 0, \quad \text{for } i = 1, 2, \ldots, n;$$
$$y_i \geq 0, \quad \text{for } i = 1, 2, \ldots, m.$$

First argue that either the optimal objective value for $D_0$ is zero or $D_0$ is unbounded.

Points $x \in \mathbf{R}^n$ and $y \in \mathbf{R}^m$ are *complementary*, with respect to $P$ and $D$, if

$$y_i \left( b_i - \sum_{j=1}^{n} a_{ij} x_j \right) = 0, \quad \text{for } i = 1, 2, \ldots, m.$$

The next two results contain the relationship between duality and complementarity.

**Weak Complementary-Slackness Theorem.** *If feasible solutions $x$ and $y$ are complementary, then $x$ and $y$ are optimal solutions.*

*Proof.* Suppose that $x$ and $y$ are feasible. Note that

$$\sum_{i=1}^{m} y_i \left( b_i - \sum_{j=1}^{n} a_{ij} x_j \right) = \sum_{i=1}^{m} y_i b_i - \sum_{j=1}^{n} \sum_{i=1}^{m} y_i a_{ij} x_j = \sum_{i=1}^{m} y_i b_i - \sum_{j=1}^{n} c_j x_j.$$

If $x$ and $y$ are complementary, then the leftmost expression in the preceding equation chain is equal to 0. Therefore, $\sum_{j=1}^{n} c_j x_j = \sum_{i=1}^{m} y_i b_i$, so, by the Weak Duality Theorem, $x$ and $y$ are optimal solutions. ■

**Strong Complementary-Slackness Theorem.** *If $x$ and $y$ are optimal solutions to $P$ and $D$, respectively, then $x$ and $y$ are complementary.*

*Proof.* Suppose that $x$ and $y$ are optimal. Then, by the Strong Duality Theorem, the rightmost expression in the equation chain of the last proof is 0. Therefore, we have

$$\sum_{i=1}^{m} y_i \left( b_i - \sum_{j=1}^{n} a_{ij} x_j \right) = 0.$$

However, by feasibility, we have $y_i(b_i - \sum_{j=1}^{n} a_{ij} x_j) \geq 0$, for $i = 1, 2, \ldots, m$. Therefore, $x$ and $y$ must be complementary. ■

The Duality and Complementary-Slackness Theorems can be used to prove a very useful characterization of optimality for linear programs over polytopes having a natural decomposition. For $k = 1, 2, \ldots, p$, let

$$\mathcal{P}_k := \left\{ x \in \mathbf{R}^n_+ : \sum_{j=1}^{n} a^k_{ij} x_j \leq b^k_i, \text{ for } i = 1, 2, \ldots, m(k) \right\},$$

and consider the linear program

$$(P) \qquad \max\left\{\sum_{j=1}^{n} c_j x_j \ : \ x \in \bigcap_{k=1}^{p} \mathcal{P}_k\right\}.$$

Suppose that the $c_j^k$ are defined so that $c_j = \sum_{k=1}^{p} c_j^k$. Such a decomposition of $c \in \mathbf{R}^n$ is called a *weight splitting* of $c$. For $k = 1, 2, \ldots, p$, consider the linear programs

$$(P_k) \qquad \max\left\{\sum_{j=1}^{n} c_j^k x_j \ : \ x \in \mathcal{P}_k\right\}.$$

**Proposition (Sufficiency of weight splitting).** *Given a weight splitting of $c$, if $\overline{x} \in \mathbf{R}^n$ is optimal for all $P_k$ ($k = 1, 2, \ldots, p$), then $\overline{x}$ is optimal for $P$.*

*Proof.* Suppose that $\overline{x}$ is optimal for all $P_k$ ($k = 1, 2, \ldots, p$). Let $\overline{y}^k$ be optimal to the dual of $P_k$:

$$(D_k) \qquad \begin{aligned} &\min \sum_{i=1}^{m(k)} y_i^k b_i^k \\ &\text{subject to:} \\ &\sum_{i=1}^{m(k)} y_i^k a_{ij}^k \ge c_j^k, \ \text{ for } j = 1, 2, \ldots, n; \\ &\qquad y_i^k \ge 0, \ \text{ for } i = 1, 2, \ldots, m(k). \end{aligned}$$

Now, $\overline{x}$ is feasible for $P$. Because we have a weight splitting of $c$, $(\overline{y}^1, \overline{y}^2, \ldots, \overline{y}^p)$ is feasible for the dual of $P$:

$$(D) \qquad \begin{aligned} &\min \sum_{k=1}^{p} \sum_{i=1}^{m(k)} y_i^k b_i^k \\ &\text{subject to:} \\ &\sum_{k=1}^{p} \sum_{i=1}^{m(k)} y_i^k a_{ij}^k \ge c_j, \ \text{ for } j = 1, 2, \ldots, n; \\ &\qquad y_i^k \ge 0, \ \text{ for } k = 1, 2, \ldots, p, \\ &\qquad\qquad i = 1, 2, \ldots, m(k). \end{aligned}$$

Optimality of $\overline{x}$ for $P_k$ and $\overline{y}^k$ for $D_k$ implies that $\sum_{j=1}^{n} c_j^k \overline{x}_j = \sum_{i=1}^{m(k)} \overline{y}_i^k b_i^k$ when the Strong Duality Theorem is applied to the pair $P_k, D_k$. Using the fact that we have a weight splitting, we can conclude that $\sum_{j=1}^{n} c_j \overline{x}_j = \sum_{k=1}^{p} \sum_{i=1}^{m(k)} \overline{y}_i^k b_i^k$. The result follows by application of the Weak Duality Theorem to the pair $P, D$. ∎

**Proposition (Necessity of weight splitting).** *If $\overline{x} \in \mathbf{R}^n$ is optimal for $P$, then there exists a weight splitting of $c$ so that $\overline{x}$ is optimal for all $P_k$ ($k = 1, 2, \ldots, p$).*

*Proof.* Suppose that $\overline{x}$ is optimal for $P$. Obviously $\overline{x}$ is feasible for $P_k$ ($k = 1, 2, \ldots, p$). Let $(\overline{y}^1, \overline{y}^2, \ldots, \overline{y}^p)$ be an optimal solution of $D$. Let $c_j^k := \sum_{i=1}^{m(k)} \overline{y}_i^k a_{ij}^k$, so $\overline{y}^k$ is feasible for $D_k$. Note that it is *not* claimed that this is a weight splitting of $c$. However, because $(\overline{y}^1, \overline{y}^2, \ldots, \overline{y}^p)$ is feasible for $D$, we do have

$$\sum_{k=1}^{p} c_j^k = \sum_{k=1}^{p} \overline{y}_i^k a_{ij}^k \geq c_j.$$

Therefore, we have a natural "weight covering" of $c$.

Applying the Weak Duality Theorem to the pair $P_k$, $D_k$ gives $\sum_{j=1}^{n} c_j^k \overline{x}_j \leq \sum_{i=1}^{m(k)} \overline{y}_i^k b_i^k$. Adding up over $k$ gives the following right-hand inequality, and the left-hand inequality follows from $\overline{x} \geq 0$ and $c \leq \sum_{k=1}^{p} c^k$:

$$\sum_{j=1}^{n} c_j \overline{x}_j \leq \sum_{k=1}^{p} \sum_{j=1}^{n} c_j^k \overline{x}_j \leq \sum_{k=1}^{p} \sum_{i=1}^{m(k)} \overline{y}_i^k b^k.$$

The Strong Duality Theorem applied to the pair $P$, $D$ implies that we have equality throughout. Therefore, we must have

$$\sum_{j=1}^{n} c_j^k \overline{x}_j = \sum_{i=1}^{m(k)} \overline{y}_i^k b^k$$

for all $k$, and, applying the Weak Duality Theorem for the pair $P_k$, $D_k$, we conclude that $\overline{x}$ is optimal for $P_k$ and $\overline{y}^k$ is optimal for $D_k$.

Now, suppose that

$$\sum_{k=1}^{p} c_j^k > c_j,$$

for some $j$. Then

$$\sum_{k=1}^{p} \overline{y}_i^k a_{ij}^k > c_j,$$

for that $j$. Applying the Weak Complementary-Slackness Theorem to the pair $P_k$, $D_k$, we can conclude that $\overline{x}_j = 0$. If we choose any $\overline{k}$ and reduce $c_j^{\overline{k}}$ until we have

$$\sum_{k=1}^{p} c_j^k = c_j,$$

we do not disturb the optimality of $\overline{x}$ for $P_k$ with $k \neq \overline{k}$. Also, because $\overline{x}_j = 0$, the Weak Complementary-Slackness Theorem applied to the pair $P_k$, $D_k$ implies that $\overline{x}$ is still optimal for $P_{\overline{k}}$. ■

## 0.3 Basic Solutions and the Primal Simplex Method

Straightforward transformations can be effected so that any linear program is brought into the *standard form*:

$$(P') \qquad \max \sum_{j=1}^{n} c_j x_j$$

subject to:

$$\sum_{j=1}^{n} a_{ij} x_j = b_i, \quad \text{for } i = 1, 2, \ldots, m;$$

$$x_j \geq 0, \quad \text{for } j = 1, 2, \ldots, n,$$

where it is assumed that the $m \times n$ matrix $A$ of constraint coefficients has full row rank.

We consider partitions of the indices $\{1, 2, \ldots, n\}$ into an ordered *basis* $\beta = (\beta_1, \beta_2, \ldots, \beta_m)$ and *nonbasis* $\eta = (\eta_1, \eta_2, \ldots, \eta_{n-m})$ so that the columns $A_{\beta_1}, A_{\beta_2}, \ldots, A_{\beta_m}$ are linearly independent. The *basic solution* $x^*$ associated with the "basic partition" $\beta, \eta$ arises if we set $x^*_{\eta_1} = x^*_{\eta_2} = \cdots = x^*_{\eta_{n-m}} = 0$ and let $x^*_{\beta_1}, x^*_{\beta_2}, \ldots, x^*_{\beta_m}$ be the unique solution to the remaining system

$$\sum_{j=1}^{m} a_{i\beta_j} x_{\beta_j} = b_i, \quad \text{for } i = 1, 2, \ldots, m.$$

In matrix notation, the basic solution $x^*$ can be expressed as $x^*_\eta = 0$ and $x^*_\beta = A_\beta^{-1} b$. If $x^*_\beta \geq 0$ then the basic solution is feasible to $P'$. Depending on whether the basic solution $x^*$ is feasible or optimal to $P'$, we may refer to the basis $\beta$ as being *primal feasible* or *primal optimal*.

The dual linear program of $P'$ is

$$(D') \qquad \min \sum_{i=1}^{m} y_i b_i$$

subject to:

$$\sum_{i=1}^{m} y_i a_{ij} \geq c_j, \quad \text{for } i = 1, 2, \ldots, n.$$

Associated with the basis $\beta$ is a potential solution to $D'$. The *basic dual solution* associated with $\beta$ is the unique solution $y_1^*, y_2^*, \ldots, y_m^*$ of

$$\sum_{i=1}^{m} y_i a_{i\beta_j} = c_{\beta_j}, \quad \text{for } j = 1, 2, \ldots, n.$$

In matrix notation, we have $y^* = c_\beta A_\beta^{-1}$. The *reduced cost* of $x_{\eta_j}$ is defined by $\overline{c}_{\eta_j} := c_{\eta_j} - \sum_{i=1}^{m} y_i^* a_{i\eta_j}$. If $\overline{c}_{\eta_j} \leq 0$, for $j = 1, 2, \ldots, m$, then the basic dual solution $y^*$is feasible to $D'$. In matrix notation this is expressed as $c_\eta - y^* A_\eta \leq 0$, or, equivalently, $\overline{c}_\eta \leq 0$. Sometimes it is convenient to write $\overline{A}_\eta := A_\beta^{-1} A_\eta$, in which case we also have $c_\eta - y^* A_\eta = c_\eta - c_\beta \overline{A}_\eta$. Depending on whether the basic dual solution $y^*$ is feasible or optimal to $D'$, we may say that $\beta$ is *dual feasible* or *dual optimal*. Finally, we say that the basis $\beta$ is *optimal* if it is both primal optimal and dual optimal.

**Weak Optimal-Basis Theorem.** *If the basis $\beta$ is both primal feasible and dual feasible, then $\beta$ is optimal.*

*Proof.* Let $x^*$ and $y^*$ be the basic solutions associated with $\beta$. We can see that

$$cx^* = c_\beta x_\beta^* = c_\beta A_\beta^{-1} b = y^* b.$$

Therefore, if $x^*$ and $y^*$ are feasible, then, by the Weak Duality Theorem, $x^*$ and $y^*$ are optimal. ∎

In fact, if $P'$ and $D'$ are feasible, then there is a basis $\beta$ that is both primal feasible and dual feasible (hence, optimal). We prove this, in a constructive manner, by specifying an algorithm. The algorithmic framework is that of a "simplex method". A convenient way to carry out a simplex method is by working with "simplex tables". The table

| $x_0$ | $x$ | rhs |
|---|---|---|
| 1 | $-c$ | 0 |
| 0 | $A$ | $b$ |

(where rhs stands for right-hand side) is just a convenient way of expressing the system of equations

$$x_0 - \sum_{j=1}^{n} c_j x_j = 0,$$

$$\sum_{j=1}^{n} a_{ij} x_j = b_i, \quad \text{for } i = 1, 2, \ldots, m.$$

Solving for the basic variables, we obtain an equivalent system of equations, which we express as the simplex table

| $x_0$ | $x_\beta$ | $x_\eta$ | rhs |
|---|---|---|---|
| 1 | 0 | $-c_\eta + c_\beta A_\beta^{-1} A_\eta$ | $c_\beta A_\beta^{-1} b$ |
| 0 | $I$ | $A_\beta^{-1} A_\eta$ | $A_\beta^{-1} b$ |

The same simplex table can be reexpressed as

| $x_0$ | $x_\beta$ | $x_\eta$ | rhs |
|---|---|---|---|
| 1 | 0 | $-\overline{c}_\eta$ | $c_\beta x_\beta^*$ |
| 0 | $I$ | $\overline{A}_\eta$ | $x_\beta^*$ |

In this form, it is very easy to check whether $\beta$ is an optimal basis. We just check that the basis $\beta$ is primal feasible (i.e., $x_\beta^* \geq 0$) and dual feasible (i.e., $\overline{c}_\eta \leq 0$).

The Primal Simplex Method is best explained by an overview that leaves a number of dangling issues that are then addressed one by one afterward. The method starts with a primal feasible basis $\beta$. *(Issue 1: Describe how to obtain an initial primal feasible basis.)*

Then, if there is some $\eta_j$ for which $\overline{c}_{\eta_j} > 0$, we choose a $\beta_i$ so that the new basis

$$\beta' := (\beta_1, \beta_2, \ldots, \beta_{i-1}, \eta_j, \beta_{i+1}, \ldots, \beta_m)$$

is primal feasible. We let the new nonbasis be

$$\eta' := (\eta_1, \eta_2, \ldots, \eta_{j-1}, \beta_i, \eta_{j+1}, \ldots, \eta_{n-m}).$$

The index $\beta_i$ is eligible to leave the basis (i.e., guarantee that $\beta'$ is primal feasible) if it satisfies the "ratio test"

$$i := \operatorname{argmin}\left\{ \frac{x_{\beta_k}^*}{\overline{a}_{k,\eta_j}} \;:\; \overline{a}_{k,\eta_j} > 0 \right\}$$

*(Issue 2: Verify the validity of the ratio test; Issue 3: Describe how the primal linear program is unbounded if there are no k for which $\overline{a}_{k,\eta_j} > 0$.)*

After the "pivot", (1) the variable $x_{\beta_i} = x_{\eta'_j}$ takes on the value 0, (2) the variable $x_{\eta_j} = x_{\beta'_i}$ takes on the value $x_{\beta'_i}^* = \frac{x_{\beta_i}^*}{\overline{a}_{i,\eta_j}}$, and (3) the increase in the objective-function value (i.e., $c_{\beta'} x_{\beta'}^* - c_\beta x_\beta^*$) is equal to $\overline{c}_{\eta_j} x_{\beta'_i}^*$. This amounts to a positive increase as long as $x_{\beta_i}^* > 0$. Then, as long as we get these positive increases at each pivot, the method must terminate since there are only a finite number of bases. *(Issue 4: Describe how to guarantee termination if $x^*_{\beta_i} = 0$ at some pivots.)*

Next, we resolve the dangling issues.

*Issue 1:* Because we can multiply some rows by $-1$, as needed, we assume, without loss of generality, that $b \geq 0$. We choose any basis $\beta$, we then introduce an "artificial variable" $x_{n+1}$, and we formulate the "Phase-I" linear program, which we encode as the table

| $x_0$ | $x_\beta$ | $x_\eta$ | $x_{n+1}$ | rhs |
|---|---|---|---|---|
| 1 | 0 | 0 | 1 | 0 |
| 0 | $A_\beta$ | $A_\eta$ | $A_\beta \mathbf{e}$ | $b$ |

This Phase-I linear program has the objective of minimizing the nonnegative variable $x_{n+1}$. Indeed, $P'$ has a feasible solution if and only if the optimal solution of this Phase-I program has $x_{n+1} = 0$.

Equivalently, we can express the Phase-I program with the simplex table

| $x_0$ | $x_\beta$ | $x_\eta$ | $x_{n+1}$ | rhs |
|---|---|---|---|---|
| 1 | 0 | 0 | 1 | 0 |
| 0 | $I$ | $A_\beta^{-1} A_\eta$ | $\mathbf{e}$ | $x^*_\beta = A_\beta^{-1} b$ |

If $x^*_\beta$ is feasible for $P'$, then we can initiate the Primal Simplex Method for $P'$, starting with the basis $\beta$. Otherwise, we choose the $i$ for which $x^*_{\beta_i}$ is most negative, and we pivot to the basis

$$\beta' := (\beta_1, \ldots, \beta_{i-1}, n+1, \beta_{i+1}, \ldots, \beta_m).$$

It is easy to check that $x^*_{\beta'} \geq 0$. Now we just apply the Primal Simplex Method to this Phase-I linear program, starting from the basis $\beta'$. In applying the Primal Simplex Method to the Phase-I program, we make one simple refinement: *When the ratio test is applied, if $n + 1$ is eligible to leave the basis, then $n + 1$ does leave the basis.*

If $n + 1$ ever leaves the basis, then we have a starting basis for applying the Primal Simplex Method to $P'$. If, at termination, we have $n + 1$ in the basis, then $P'$ is infeasible. (It is left to the reader to check that our refinement of the ratio test ensures that we cannot terminate with a basis $\beta$ containing $n + 1$ and having $x_{n+1} = 0$ in the associated basic solution.)

*Issue 2:* Because

$$\begin{aligned} A_{\beta'} &= [A_{\beta_1}, \ldots, A_{\beta_{i-1}}, A_{\eta_j}, A_{\beta_{i-1}}, \ldots, A_{\beta_i}] \\ &= A_\beta[\mathbf{e}^1, \ldots, \mathbf{e}^{i-1}, \overline{A}_{\eta_j}, \mathbf{e}^{i+1}, \ldots, \mathbf{e}^m], \end{aligned}$$

we have

$$A_{\beta'}^{-1} = [\mathbf{e}^1, \ldots, \mathbf{e}^{i-1}, \overline{A}_{\eta_j}, \mathbf{e}^{i+1}, \ldots, \mathbf{e}^m]^{-1} A_\beta^{-1},$$

and

$$\begin{aligned} x^*_{\beta'} &= A_{\beta'}^{-1} b \\ &= [\mathbf{e}^1, \ldots, \mathbf{e}^{i-1}, \overline{A}_{\eta_j}, \mathbf{e}^{i+1}, \ldots, \mathbf{e}^m]^{-1} A_\beta^{-1} b \\ &= [\mathbf{e}^1, \ldots, \mathbf{e}^{i-1}, \overline{A}_{\eta_j}, \mathbf{e}^{i+1}, \ldots, \mathbf{e}^m]^{-1} x^*_\beta. \end{aligned}$$

Now, the inverse of

$$[\mathbf{e}^1, \ldots, \mathbf{e}^{i-1}, \overline{A}_{\eta_j}, \mathbf{e}^{i+1}, \ldots, \mathbf{e}^m]$$

is

$$\left[\begin{array}{c|c|c|c|c|c|c} & & & -\dfrac{\overline{a}_{1,\eta_j}}{\overline{a}_{i,\eta_j}} & & & \\ & & & \vdots & & & \\ & & & -\dfrac{\overline{a}_{i-1,\eta_j}}{\overline{a}_{i,\eta_j}} & & & \\ \mathbf{e}^1 & \ldots & \mathbf{e}^{i-1} & \dfrac{1}{\overline{a}_{i,\eta_j}} & \mathbf{e}^{i+1} & \ldots & \mathbf{e}^m \\ & & & -\dfrac{\overline{a}_{i+1,\eta_j}}{\overline{a}_{i,\eta_j}} & & & \\ & & & \vdots & & & \\ & & & -\dfrac{\overline{a}_{m,\eta_j}}{\overline{a}_{i,\eta_j}} & & & \end{array}\right].$$

Therefore,

$$x^*_{\beta'_k} = \begin{cases} x^*_{\beta_k} - \frac{\overline{a}_{k,\eta_j}}{\overline{a}_{i,\eta_j}} x^*_{\beta_i}, & \text{for } k \neq i \\ \frac{x^*_{\beta_i}}{\overline{a}_{i,\eta_j}}, & \text{for } k = i \end{cases}.$$

To ensure that $x^*_{\beta'_i} \geq 0$, we choose $i$ so that $\overline{a}_{i,\eta_j} > 0$. To ensure that $x^*_{\beta'_k} \geq 0$, for $k \neq i$, we then need

$$\frac{x^*_{\beta_i}}{\overline{a}_{i,\eta_j}} \leq \frac{x^*_{\beta_k}}{\overline{a}_{k,\eta_j}}, \text{ for } k \neq i, \text{ such that } \overline{a}_{k,\eta_j} > 0.$$

*Issue 3:* We consider the solutions $\widehat{x} := x^* + \epsilon h$, where $h \in \mathbf{R}^n$ is defined by $h_\eta := \mathbf{e}^j$ and $h_\beta := -\overline{A}_{\eta_j}$. We have

$$A\widehat{x} = Ax + \epsilon Ah = b + \epsilon \left(A_{\eta_j} - A_\beta \overline{A}_{\eta_j}\right) = b.$$

If we choose $\epsilon \geq 0$, then $\widehat{x}_\beta = x_\beta - \epsilon \overline{A}_{\eta_j}$, which is nonnegative because we are assuming that $\overline{A}_{\eta_j} \leq 0$. Finally, $\widehat{x}_\eta = x_\eta + \epsilon h_\eta = \epsilon \mathbf{e}^j$, which is also nonnegative when $\epsilon \geq 0$. Therefore, $\widehat{x}$ is feasible for all $\epsilon \geq 0$. Now, considering the objective value, we have $c\widehat{x} = cx^* + \epsilon ch = cx^* + \epsilon \overline{c}_{\eta_j}$. Therefore, by choosing $\epsilon$ large, we can make $c\widehat{x}$ as large as we like.

*Issue 4:* We will work with polynomials (of degree no more than $m$) in $\epsilon$, where we consider $\epsilon$ to be an arbitrarily small indeterminant. In what follows, we write $\leq_\epsilon$ to denote the induced ordering of polynomials. An important point is that if $p(\epsilon) \leq_\epsilon q(\epsilon)$, then $p(0) \leq q(0)$.

We algebraically perturb the right-hand-side vector $b$ by replacing each $b_i$ with $b_i + \epsilon^i$, for $i = 1, 2, \ldots, m$. We carry out the Primal Simplex Method with the understanding that, in applying the ratio test, we use the $\leq_\epsilon$ ordering. For any basis $\beta$, the value of the basic variables $x^*_{\beta_i}$ is the polynomial

$$x^*_{\beta_i} = A_\beta^{-1} b + \sum_{k=1}^{m} \left(A_\beta^{-1}\right)_{ik} \epsilon^k.$$

We cannot have $x^*_{\beta_i}$ equal to the zero polynomial, as that would imply that the $i$th row of $A_\beta^{-1}$ is all zero – contradicting the nonsingularity of $A_\beta$. Therefore, the objective increase is always positive with respect to the $\leq_\epsilon$ ordering. Therefore, we will never revisit a basis.

Furthermore, as the Primal Simplex Method proceeds, the ratio test seeks to maintain the nonnegativity of the basic variables with respect to the $\leq_\epsilon$ ordering. However, this implies ordinary nonnegativity, evaluating the polynomials at 0. Therefore, each pivot yields a feasible basic solution for the unperturbed problem $P'$. Therefore, we really are carrying out valid steps of the Primal Simplex Method with respect to the unperturbed problem $P'$.

By filling in all of these details, we have provided a constructive proof of the following result.

**Strong Optimal-Basis Theorem.** *If $P'$ and $D'$ are feasible, then there is a basis $\beta$ that is both primal feasible and dual feasible (hence, optimal).*

This section closes with a geometric view of the feasible basic solutions visited by the Primal Simplex Method. An *extreme point* of a convex set $C$ is a point $x \in C$, such that $x^1, x^2 \in C, 0 < \lambda < 1, x = \lambda x^1 + (1-\lambda)x^2$, implies $x = x^1 = x^2$. That is, $x$ is extreme if it is not an interior point of a closed line segment contained in $C$.

**Theorem (Feasible basic solutions and extreme points).** *The set of feasible basic solutions of $P'$ is the same as the set of extreme points of $\mathcal{P}'$.*

*Proof.* Let $\widetilde{x}$ be a feasible basic solution of $P'$, with corresponding basis $\beta$ and nonbasis $\eta$. Suppose that $\widetilde{x} = \lambda x^1 + (1-\lambda)x^2$, with $x^1, x^2 \in \mathcal{P}'$, and $0 < \lambda < 1$. Because $x_\eta = 0, x^1_\eta, x^2_\eta \geq 0, \widetilde{x}_\eta = \lambda x^1_\eta + (1-\lambda)x^2_\eta$, we must have $x^1_\eta = x^2_\eta = \widetilde{x}_\eta = 0$. Then we must have $A_\beta x^l_\beta = b$, for $l = 1, 2$. However, this system has the unique solution $\widetilde{x}_\beta$. Therefore, $x^1_\beta = x^2_\beta = \widetilde{x}_\beta$. Hence, $x^1 = x^2 = \widetilde{x}$.

Conversely, suppose that $\widetilde{x}$ is an extreme point of $\mathcal{P}'$. Let

$$\phi := \left\{ j \in \{1, 2, \ldots, n\} \ : \ \widetilde{x}_j > 0 \right\}.$$

We claim that the columns of $A_\phi$ are linearly independent. To demonstrate this, we suppose otherwise. Then there is a vector $w_\phi \in \mathbf{R}^{|\phi|}$, such that $w_\phi$ is not the zero vector, and $A_\phi w_\phi = 0$. Next, we extend $w_\phi$ to $w \in \mathbf{R}^n$ by letting $w_j = 0$, for $j \notin \phi$. Therefore, we have $Aw = 0$. Now, we let $x^1 := \widetilde{x} + \epsilon w$, and $x^2 := \widetilde{x} - \epsilon w$, where $\epsilon$ is a small positive scalar. For any $\epsilon$, it is easy to check that $\widetilde{x} = \frac{1}{2}x^1 + \frac{1}{2}x^2$, $Ax^1 = b$, and $Ax^2 = b$. Moreover, for $\epsilon$ small enough, we have $x^1, x^2 \geq 0$ (by the choice of $\phi$). Therefore, $\widetilde{x}$ is not an extreme point of $\mathcal{P}'$. This contradiction establishes that the columns of $A_\phi$ are linearly independent.

Now, in an arbitrary manner, we complete $\phi$ to a basis $\beta$, and $\eta$ consists of the remaining indices. We claim that $\widetilde{x}$ is the basic solution associated with this choice of basis $\beta$. Clearly, by the choice of $\phi$, we have $x_\eta = 0$. The remaining system, $A_\beta x_\beta = b$, has a unique solution, as $A_\beta$ is nonsingular. That unique solution is $\widetilde{x}_\beta$, because $A_\beta \widetilde{x}_\beta = A_\phi \widetilde{x}_\phi = A\widetilde{x} = b$. ∎

## 0.4 Sensitivity Analysis

The optimal objective value of a linear program behaves very nicely as a function of its objective coefficients. Assume that $P$ is feasible, and define the *objective value function* $g : \mathbf{R}^n \mapsto \mathbf{R}$ by

$$g(c_1, c_2, \ldots, c_n) := \max \sum_{j=1}^{n} c_j x_j$$

$$\text{subject to:}$$

$$\sum_{j=1}^{n} a_{ij} x_j \leq b_i, \quad \text{for } i = 1, 2, \ldots, m,$$

whenever the optimal value exists (i.e., whenever $P$ is not unbounded). Using the Weak Duality Theorem, we find that the domain of $g$ is just the set of $c \in \mathbf{R}^n$ for which the following system is feasible:

$$\sum_{i=1}^{m} y_i a_{ij} = c_j, \quad \text{for } i = 1, 2, \ldots, n;$$

$$y_i \geq 0, \quad \text{for } i = 1, 2, \ldots, m.$$

Thinking of the $c_j$ as variables, we can eliminate the $y_i$ by using Fourier–Motzkin Elimination. In this way, we observe that the domain of $g$ is the solution set of a finite system of linear inequalities (in the variables $c_j$). In particular, the domain of $g$ is a convex set.

Similarly, we assume that $D$ is feasible and define the *right-hand-side value function* $f : \mathbf{R}^m \mapsto \mathbf{R}$ by

$$f(b_1, b_2, \ldots, b_m) := \max \sum_{j=1}^{n} c_j x_j$$

$$\text{subject to:}$$

$$\sum_{j=1}^{n} a_{ij} x_j \leq b_i, \quad \text{for } i = 1, 2, \ldots, m,$$

whenever the optimal value exists (i.e., whenever $P$ is feasible). As previously, we observe that the domain of $f$ is the solution set of a finite system of linear inequalities (in the variables $b_i$).

A *piecewise-linear* function is the point-wise maximum or minimum of a finite number of linear functions.

**Problem (Sensitivity Theorem)**

a. Prove that a piecewise-linear function that is the point-wise maximum (respectively, minimum) of a finite number of linear functions is convex (respectively, concave).
b. Prove that the function $g$ is convex and piecewise linear on its domain.
c. Prove that the function $f$ is concave and piecewise linear on its domain.

## 0.5 Polytopes

The *dimension* of a polytope $\mathcal{P}$, denoted $\dim(\mathcal{P})$, is one less than the maximum number of affinely independent points in $\mathcal{P}$. Therefore, the empty set has dimension $-1$. The polytope $\mathcal{P} \subset \mathbf{R}^n$ is *full dimensional* if $\dim(\mathcal{P}) = n$. The linear equations

$$\sum_{j=1}^{n} \alpha_j^i x_j = \beta_i, \text{ for } i = 1, 2, \ldots, m,$$

are *linearly independent* if the points $\binom{\alpha^i}{\beta_i} \in \mathbf{R}^{n+1}$, $i = 1, 2, \ldots, m$ are linearly independent.

**Dimension Theorem.** $\dim(\mathcal{P})$ *is* $n$ *minus the maximum number of linearly independent equations satisfied by all points of* $\mathcal{P}$.

*Proof.* Let $\mathcal{P} := \text{conv}(X)$. For $x \in X$, let $\tilde{x} := \binom{x}{1} \in \mathbf{R}^{n+1}$. Arrange the points $\tilde{x}$ as the rows of a matrix $G$ with $n+1$ columns. We have that $\dim(\mathcal{P}) + 1 = \dim(\text{r.s.}(G))$. Then, by using the rank-nullity theorem, we have that $\dim(\mathcal{P}) + 1 = (n+1) - \dim(\text{n.s.}(G))$. The result follows by noting that, for $\alpha \in \mathbf{R}^n$, $\beta \in \mathbf{R}$, we have $\binom{\alpha}{\beta} \in \text{n.s.}(G)$ if and only if $\sum_{j=1}^n \alpha_j x_j = \beta$ for all $x \in \mathcal{P}$. ■

An inequality $\sum_{j=1}^n \alpha_j x_j \le \beta$ is *valid* for the polytope $\mathcal{P}$ if every point in $\mathcal{P}$ satisfies the inequality. The valid inequality $\sum_{j=1}^n \alpha_j x_j \le \beta$ *describes* the *face*

$$\mathcal{F} := P \cap \left\{ x \in \mathbf{R} : \sum_{j=1}^{n} \alpha_j x_j = \beta \right\}.$$

It is immediate that if $\sum_{j=1}^n \alpha_j x_j \le \beta$ describes the nonempty face $\mathcal{F}$ of $\mathcal{P}$, then $x^* \in \mathcal{P}$ maximizes $\sum_{j=1}^n \alpha_j x_j$ over $\mathcal{P}$ if and only if $x^* \in \mathcal{F}$. Furthermore, if $\mathcal{F}$ is a face of

$$\mathcal{P} = \left\{ x \in \mathbf{R}^n : \sum_{j=1}^{n} a_{ij} x_j \le b_i, \text{ for } i = 1, 2, \ldots, m \right\},$$

then

$$\mathcal{F} = \mathcal{P} \cap \left\{ x \in \mathbf{R}^n : \sum_{j=1}^{n} a_{i(k),j} x_j = b_i, \text{ for } k = 1, 2, \ldots, r \right\},$$

where $\{i(1), i(2), \ldots, i(r)\}$ is a subset of $\{1, 2, \ldots, m\}$. Hence, $\mathcal{P}$ has a finite number of faces.

Faces of polytopes are themselves polytopes. Every polytope has the empty set and itself as *trivial* faces. Faces of dimension 0 are called *extreme points*. Faces of dimension $\dim(\mathcal{P}) - 1$ are called *facets*.

A partial converse to Weyl's Theorem can be established:

**Minkowski's Theorem (for polytopes).** *If*

$$\mathcal{P} = \left\{ x \in \mathbf{R}^n \ : \ \sum_{j=1}^{n} a_{ij} x_j \leq b_i, \ \textit{for}\ i = 1, 2, \ldots, m \right\},$$

*and $\mathcal{P}$ is bounded, then $\mathcal{P}$ is a polytope.*

*Proof.* Let $X$ be the set of extreme points of $\mathcal{P}$. Clearly $\mathrm{conv}(X) \subset \mathcal{P}$. Suppose that $\widetilde{x} \in \mathcal{P} \setminus \mathrm{conv}(X)$. Then there fail to exist $\lambda_x$, $x \in X$, such that

$$\begin{aligned} \widetilde{x}_j &= \sum_{x \in X} \lambda_x x_j, \text{ for } j = 1, 2, \ldots, n; \\ 1 &= \sum_{x \in X} \lambda_x; \\ \lambda_x &\geq 0, \quad \forall \quad x \in X. \end{aligned}$$

By the Theorem of the Alternative for Linear Inequalities, there exist $c \in \mathbf{R}^n$ and $t \in \mathbf{R}$ such that

$$\begin{aligned} t + \sum_{j=1}^{n} c_j x_j &\geq 0, \quad \forall \quad x \in X; \\ t + \sum_{j=1}^{n} c_j \widetilde{x}_j &< 0. \end{aligned}$$

Therefore, we have that $\widetilde{x}$ is a feasible solution to the linear program

$$\begin{aligned} &\min \sum_{j=1}^{n} c_j x_j \\ &\text{subject to:} \\ &\sum_{j=1}^{n} a_{ij} x_j \leq b_i, \quad \text{for } i = 1, 2, \ldots, m, \end{aligned}$$

and it has objective value less than $-t$, but all extreme points of the feasible region have objective value at least $t$. Therefore, no extreme point solves the linear program. This is a contradiction. ■

**Redundancy Theorem.** *Valid inequalities that describe faces of $\mathcal{P}$ having dimension less than* $\dim(\mathcal{P}) - 1$ *are redundant.*

*Proof.* Without loss of generality, we can describe $\mathcal{P}$ as the solution set of

$$\sum_{j=1}^{n} \widehat{a}_{ij} x_j = \widehat{b}_i, \text{ for } i = 1, 2, \ldots, k;$$

$$\sum_{j=1}^{n} a_{ij} x_j \leq b_i, \text{ for } i = 0, 1, \ldots, m,$$

where the equations

$$\sum_{j=1}^{n} \widehat{a}_{ij} x_j = \widehat{b}_i, \text{ for } i = 1, 2, \ldots, k,$$

are linearly independent, and such that for $i = 0, 1, \ldots, m$, there exist points $\widetilde{x}^i$ in $\mathcal{P}$ with

$$\sum_{j=1}^{n} a_{ij} \widetilde{x}^i_j < b_i.$$

With these assumptions, it is clear that $\dim(\mathcal{P}) = n - k$.

Let

$$\widetilde{x} := \sum_{i=0}^{m} \frac{1}{m+1} \widetilde{x}^i.$$

We have

$$\sum_{j=1}^{n} \widehat{a}_{ij} \widetilde{x}_j = \widehat{b}_i, \text{ for } i = 1, 2, \ldots, k;$$

$$\sum_{j=1}^{n} a_{ij} \widetilde{x}_j < b_i, \text{ for } i = 0, 1, \ldots, m.$$

Therefore, the point $\widetilde{x}$ is in the relative interior of $\mathcal{P}$.

Without loss of generality, consider the face $\mathcal{F}$ described by

$$\sum_{j=1}^{n} a_{0j} x_j \leq b_0.$$

We have that $\dim(\mathcal{F}) \leq \dim(\mathcal{P}) - 1$. Suppose that this inequality is necessary for describing $\mathcal{P}$. Then there is a point $x^1$ such that

$$\sum_{j=1}^{n} \widehat{a}_{ij} x_j^1 = \widehat{b}_i, \text{ for } i = 1, 2, \ldots, k;$$

$$\sum_{j=1}^{n} a_{0j} x_j^1 > b_0;$$

$$\sum_{j=1}^{n} a_{ij} x_j^1 \leq b_i, \text{ for } i = 1, 2, \ldots, m.$$

It follows that on the line segment between $\widetilde{x}$ and $x^1$ there is a point $x^2$ such that

$$\sum_{j=1}^{n} \widehat{a}_{ij} x_j^2 = \widehat{b}_i, \text{ for } i = 1, 2, \ldots, k;$$

$$\sum_{j=1}^{n} a_{0j} x_j^2 = b_0;$$

$$\sum_{j=1}^{n} a_{ij} x_j^2 < b_i, \text{ for } i = 1, 2, \ldots, m.$$

This point $x^2$ is in $\mathcal{F}$. Therefore, $\dim(\mathcal{F}) \geq \dim(\mathcal{P}) - 1$. Hence, $\mathcal{F}$ is a facet of $\mathcal{P}$. ■

**Theorem (Necessity of facets).** *Suppose that polytope $\mathcal{P}$ is described by a linear system. Then for each facet $\mathcal{F}$ of $\mathcal{P}$, it is necessary that some valid inequality that describes $\mathcal{F}$ be used in the description of $\mathcal{P}$.*

*Proof.* Without loss of generality, we can describe $\mathcal{P}$ as the solution set of

$$\sum_{j=1}^{n} \widehat{a}_{ij} x_j = \widehat{b}_i, \text{ for } i = 1, 2, \ldots, k;$$

$$\sum_{j=1}^{n} a_{ij} x_j \leq b_i, \text{ for } i = 1, 2, \ldots, m,$$

where the equations

$$\sum_{j=1}^{n} \widehat{a}_{ij} x_j = \widehat{b}_i, \text{ for } i = 1, 2, \ldots, k,$$

are linearly independent, and such that for $i = 1, 2, \ldots, m$, there exist points

$\widetilde{x}^i$ in $\mathcal{P}$ with

$$\sum_{j=1}^{n} a_{ij}\widetilde{x}^i_j < b_i.$$

Suppose that $\mathcal{F}$ is a facet of $\mathcal{P}$, but no inequality describing $\mathcal{F}$ appears in the preceding description. Suppose that

$$\sum_{j=1}^{n} a_{0j}x_j \le b_0$$

describes $\mathcal{F}$. Let $\widetilde{x}$ be a point in the relative interior of $\mathcal{F}$. Certainly there is no nontrivial solution $y \in \mathbf{R}^k$ to

$$\sum_{i=1}^{k} y_i\widehat{a}_{ij} = 0, \text{ for } j = 1, 2, \ldots, n.$$

Therefore, there is a solution $z \in \mathbf{R}^n$ to

$$\sum_{j=1}^{n} \widehat{a}_{ij}z_j = 0, \text{ for } i = 1, 2, \ldots, k;$$

$$\sum_{j=1}^{n} a_{0j}z_j > 0.$$

Now, for small enough $\epsilon > 0$, we have $\widetilde{x} + \epsilon z \in \mathcal{P}$, but $\widetilde{x} + \epsilon z$ violates the inequality

$$\sum_{j=1}^{n} a_{0j}x_j \le b_0$$

describing $\mathcal{F}$ (for all $\epsilon > 0$). ∎

Because of the following result, it is easier to work with full-dimensional polytopes.

**Unique Description Theorem.** *Let $\mathcal{P}$ be a full-dimensional polytope. Then each valid inequality that describes a facet of $\mathcal{P}$ is unique, up to multiplication by a positive scalar. Conversely, if a face $\mathcal{F}$ is described by a unique inequality, up to multiplication by a positive scalar, then $\mathcal{F}$ is a facet of $\mathcal{P}$.*

*Proof.* Let $\mathcal{F}$ be a nontrivial face of a full-dimensional polytope $\mathcal{P}$. Suppose that $\mathcal{F}$ is described by

$$\sum_{j=1}^{n} \alpha_j^1 x_j \le \beta^1$$

and by

$$\sum_{j=1}^{n} \alpha_j^2 x_j \le \beta^2,$$

with $\binom{\alpha^1}{\beta^1} \neq \lambda \binom{\alpha^2}{\beta^2}$ for any $\lambda > 0$. It is certainly the case that $\alpha^1 \neq 0$ and $\alpha^2 \neq 0$, as $\mathcal{F}$ is a nontrivial face of $\mathcal{P}$. Furthermore, $\binom{\alpha^1}{\beta^1} \neq \lambda \binom{\alpha^2}{\beta^2}$ for any $\lambda < 0$, because if that were the case we would have $\mathcal{P} \subset \sum_{j=1}^{n} \alpha_j^1 x_j = \beta^1$, which is ruled out by the full dimensionality of $\mathcal{P}$. We can conclude that

$$\left\{ \begin{pmatrix} \alpha^1 \\ \beta^1 \end{pmatrix}, \begin{pmatrix} \alpha^2 \\ \beta^2 \end{pmatrix} \right\}$$

is a linearly independent set. Therefore, by the Dimension Theorem, $\dim(\mathcal{F}) \le n - 2$. Hence, $\mathcal{F}$ can not be a facet of $\mathcal{P}$.

For the converse, suppose that $\mathcal{F}$ is described by

$$\sum_{j=1}^{n} \alpha_j x_j \le \beta.$$

Because $\mathcal{F}$ is nontrivial, we can assume that $\alpha \neq 0$. If $\mathcal{F}$ is not a facet, then there exists $\binom{\alpha'}{\beta'}$, with $\alpha' \neq 0$, such that $\binom{\alpha'}{\beta'} \neq \lambda \binom{\alpha}{\beta}$ for all $\lambda \neq 0$, and

$$\mathcal{F} \subset \left\{ x \in \mathbf{R}^n \; : \; \sum_{j=1}^{n} \alpha'_j x_j = \beta' \right\}.$$

Consider the inequality

$$(*) \qquad \sum_{j=1}^{n} (\alpha_j + \epsilon \alpha'_j) x_j \le \beta + \epsilon \beta',$$

where $\epsilon$ is to be determined. It is trivial to check that $(*)$ is satisfied for all $x \in \mathcal{F}$. To see that $(*)$ describes $\mathcal{F}$, we need to find $\epsilon$ so that strict inequality holds in $(*)$ for all $\hat{x} \in \mathcal{P} \setminus F$. In fact, we need to check this only for such $\hat{x}$ that are extreme points of $\mathcal{P}$. Because there are only a finite number of such $\hat{x}$, there exists $\gamma > 0$ so that

$$\sum_{j=1}^{n} \alpha_j \hat{x}_j < \beta - \gamma,$$

for all such $\hat{x}$. Therefore, it suffices to choose $\epsilon$ so that

$$\epsilon\left(\sum_{j=1}^{n}\alpha'_j\hat{x}_j - \beta'\right) \leq \gamma,$$

for all such $\hat{x}$. Because there are only a finite number of such $\hat{x}$, it is clear that we can choose $\epsilon$ appropriately. ■

## 0.6 Lagrangian Relaxation

Let $f : \mathbf{R}^m \mapsto \mathbf{R}$ be a convex function. A vector $\widetilde{h} \in \mathbf{R}^m$ is a *subgradient* of $f$ at $\widetilde{\pi}$ if

$$f(\pi) \geq f(\widetilde{\pi}) + \sum_{i=1}^{m}(\pi_i - \widetilde{\pi}_i)\widetilde{h}_i, \quad \forall \quad \pi \in \mathbf{R}^m;$$

that is, using the linear function $f(\widetilde{\pi}) + \sum_{i=1}^{m}(\pi_i - \widetilde{\pi}_i)\widetilde{h}_i$ to extrapolate from $\widetilde{\pi}$, we never overestimate $f$. The existence of a subgradient characterizes convex functions. If $f$ is differentiable at $\widetilde{\pi}$, then the subgradient of $f$ at $\widetilde{\pi}$ is unique, and it is the gradient of $f$ at $\widetilde{\pi}$.

Our goal is to minimize $f$ on $\mathbf{R}^m$. If a subgradient $\widetilde{h}$ is available for every $\widetilde{\pi} \in \mathbf{R}^m$, then we can utilize the following method for minimizing $f$.

**The Subgradient Method**

1. Let $k := 0$ and choose $\widetilde{\pi}^0 \in \mathbf{R}^m$.
2. Compute the subgradient $\widetilde{h}^k$ at $\widetilde{\pi}^k$.
3. Select a positive scalar ("step size") $\lambda_k$ .
4. Let $\widetilde{\pi}^{k+1} := \widetilde{\pi}^k - \lambda_k\widetilde{h}^k$.
5. Go to step 2 unless $\widetilde{h} = 0$ or a convergence criterion is met.

Before a description of how to choose the step size is given, we explain the form of the iteration $\widetilde{\pi}^{k+1} := \widetilde{\pi}^k - \lambda_k\widetilde{h}^k$. The idea is that, if $\lambda_k > 0$ is chosen to be quite small, then, considering the definition of the subgradient, we can hope that

$$f(\widetilde{\pi}^{k+1}) \approx f(\widetilde{\pi}^k) + \sum_{i=1}^{m}(\widetilde{\pi}_i^{k+1} - \widetilde{\pi}_i^k)\widetilde{h}_i^k.$$

Substituting $\widetilde{\pi}^k - \lambda_k\widetilde{h}^k$ for $\widetilde{\pi}^{k+1}$ on the right-hand side, we obtain

$$f(\widetilde{\pi}^{k+1}) \approx f(\widetilde{\pi}^k) - \lambda\sum_{i=1}^{m}\|\widetilde{h}^k\|^2.$$

Therefore, we can hope for a decrease in $f$ as we iterate.

As far as the convergence criterion goes, it is known that any choice of $\lambda_k > 0$ satisfying $\lim_{k\to\infty} \lambda_k = 0$ and $\sum_{k=0}^{\infty} \lambda_k = \infty$ leads to convergence, although this result does not guide practice. Also, in regard to Step 5 of the Subgradient Method, we note that, if 0 is a subgradient of $f$ at $\widetilde{\pi}$, then $\widetilde{\pi}$ minimizes $f$.

It is worth mentioning that the sequence of $f(\widetilde{\pi}^k)$ may not be nondecreasing. In many practical uses, we are satisfied with a good upper bound on the true minimum, and we may stop the iterations when $k$ reaches some value $K$ and use $\min_{k=0}^{K} f(\widetilde{\pi}^k)$ as our best upper bound on the true minimum.

Sometimes we are interested in minimizing $f$ on a convex set $C \subset \mathbf{R}^m$. In this case, in Step 1 of the Subgradient Method, we take care to choose $\widetilde{\pi}^0 \in C$. In Steps 4 and 5, we replace the subgradient $\widetilde{h}^k$ with its projection onto $C$. An important example is when $C = \{\pi \in \mathbf{R}^m \ : \ \pi \geq 0\}$. In this case the projection amounts to replacing the negative components of $\widetilde{h}^k$ with 0.

Next, we move to an important use of subgradient optimization. Let $\mathcal{P}$ be a polyhedron in $\mathbf{R}^n$. We consider the linear program

$$(P) \qquad \begin{aligned} z := \max \sum_{j=1}^{n} c_j x_j & \\ \text{subject to:} & \\ \sum_{j=1}^{n} a_{ij} x_j \leq b_i, & \quad \text{for } i = 1, 2, \ldots, m, \\ x \in \mathcal{P}. & \end{aligned}$$

For any nonnegative $\pi \in \mathbf{R}^m$, we also consider the related linear program

$$(L(\pi)) \qquad \begin{aligned} f(\pi) := \sum_{i=1}^{m} \pi_i b_i + \max \sum_{j=1}^{n} \left( c_j - \sum_{i=1}^{m} \pi_i a_{ij} \right) x_j & \\ \text{subject to:} & \\ x \in \mathcal{P}, & \end{aligned}$$

which is referred to as a *Lagrangian relaxation* of $P$.

Let $C := \{\pi \in \mathbf{R}^m \ : \ \pi \geq 0\}$.

**Lagrangian Relaxation Theorem**

a. *For all* $\pi \in C$, $z \leq f(\pi)$. *Moreover, if the* $\pi_i^*$ *are the optimal dual variables for the constraints* $\sum_{j=1}^{n} a_{ij} x_j \leq b_i$, $i = 1, 2, \ldots, m$, *in* $P$, *then* $f(\pi^*) = z$.

b. *The function* $f$ *is convex and piecewise linear on* $C$.

c. *If* $\widetilde{\pi} \in C$ *and* $\widetilde{x}$ *is an optimal solution of* $L(\widetilde{\pi})$, *then* $\widetilde{h} \in \mathbf{R}^m$, *defined by*

$$\widetilde{h}_i := b_i - \sum_{j=1}^{n} a_{ij}\widetilde{x}_j, \text{ for } i = 1, 2, \ldots, m,$$

*is a subgradient of* $f$ *at* $\widetilde{\pi}$.

*Proof*

a. If we write $L(\pi)$ in the equivalent form

$$f(\pi) := \max \sum_{j=1}^{n} c_j x_j + \sum_{i=1}^{m} \pi_i \left( b_i - \sum_{j=1}^{n} a_{ij} x_j \right)$$

subject to:

$$x \in \mathcal{P},$$

we see that every feasible solution $x$ to $P$ is also feasible to $L(\pi)$, and the objective value of such an $x$ is at least as great in $L(\pi)$ as it is in $P$.

Moreover, suppose that $\mathcal{P} = \{x \in \mathbf{R}^n : \sum_{j=1}^{n} d_{lj} x_j \leq d_l,\ l = 1, 2, \ldots, r\}$, and that $\pi^*$ and $y^*$ form the optimal solution of the dual of $P$:

$$z := \min \sum_{i=1}^{m} \pi_i b_i + \sum_{l=1}^{r} y_l d_l$$

subject to:

$$(D) \qquad \sum_{i=1}^{n} a_{ij}\pi_i + \sum_{l=1}^{r} d_{lj} y_l = c_j, \quad \text{for } j = 1, 2, \ldots, n,$$

$$\pi_i \geq 0, \quad \text{for } i = 1, 2, \ldots, m;$$

$$y_l \geq 0, \quad \text{for } l = 1, 2, \ldots, r.$$

It follows that

$$\begin{aligned} f(\pi^*) &= \sum_{i=1}^{m} \pi_i^* b_i + \max \left\{ \sum_{j=1}^{n} \left( c_j - \sum_{i=1}^{m} \pi_i^* a_{ij} \right) x_j \ : \ x \in \mathcal{P} \right\} \\ &= \sum_{i=1}^{m} \pi_i^* b_i + \min \left\{ \sum_{l=1}^{r} y_l d_l \ : \ \sum_{l=1}^{r} y_l d_{lj} = c_j - \sum_{i=1}^{m} \pi_i^* a_{ij}, \right. \\ &\qquad \left. j = 1, 2, \ldots, n;\ y_l \geq 0,\ l = 1, 2, \ldots, r \right\} \\ &\leq \sum_{i=1}^{m} \pi_i^* b_i + \sum_{l=1}^{r} y_l^* d_l \\ &= z. \end{aligned}$$

Therefore, $f(\pi^*) = z$.

b. This is left to the reader (use the same technique as was used for the Sensitivity Theorem Problem, part b, p. 28).

c.

$$
\begin{aligned}
f(\pi) &= \sum_{i=1}^{m} \pi_i b_i + \max_{x \in \mathcal{P}} \sum_{j=1}^{n} \left( c_j - \sum_{i=1}^{m} \pi_i a_{ij} \right) x_j \\
&\geq \sum_{i=1}^{m} \pi_i b_i + \sum_{j=1}^{n} \left( c_j - \sum_{i=1}^{m} \pi_i a_{ij} \right) \widetilde{x}_j \\
&= \sum_{i=1}^{m} \widetilde{\pi}_i b_i + \sum_{j=1}^{n} \left( c_j - \sum_{i=1}^{m} \widetilde{\pi}_i a_{ij} \right) \widetilde{x}_j + \sum_{i=1}^{p} (\pi_i - \widetilde{\pi}_i) \left( b_i - \sum_{j=1}^{n} a_{ij} \widetilde{x}_j \right) \\
&= f(\widetilde{\pi}) + \sum_{i=1}^{m} (\pi_i - \widetilde{\pi}_i) \widetilde{h}_i .
\end{aligned}
$$

■

This theorem provides us with a practical scheme for determining a good upper bound on $z$. We simply seek to minimize the convex function $f$ on the convex set $C$ (as indicated in part a of the theorem, the minimum value is $z$). Rather than explicitly solve the linear program $P$, we apply the Subgradient Method and repeatedly solve instances of $L(\pi)$. We note that

1. the projection operation here is quite easy – just replace the negative components of $\widetilde{h}$ with 0;
2. the constraint set for $L(\pi)$ may be much easier to work with than the constraint set of $P$;
3. reoptimizing $L(\pi)$, at each subgradient step, may not be so time consuming once the iterates $\widetilde{\pi}^k$ start to converge.

**Example (Lagrangian Relaxation).** We consider the linear program

$$
(P) \qquad
\begin{aligned}
z = \max\ & x_1 + x_2 \\
\text{subject to:} & \\
& 3x_1 - x_2 \leq 1; \\
& x_1 + 3x_2 \leq 2; \\
& 0 \leq x_1, x_2 \leq 1.
\end{aligned}
$$

Taking $\mathcal{P} := \{(x_1, x_2) \ : \ 0 \leq x_1, x_2 \leq 1\}$, we are led to the Lagrangian relaxation

$$
(L(\pi_1, \pi_2)) \qquad
\begin{aligned}
f(\pi_1, \pi_2) = \pi_1 + 2\pi_2 + \max\ & (1 - 3\pi_1 - \pi_2)x_1 + (1 + \pi_1 - 3\pi_2)x_2 \\
\text{subject to:} & \\
& 0 \leq x_1, x_2 \leq 1.
\end{aligned}
$$

The extreme points of $\mathcal{P}$ are just the corners of the standard unit square. For any point $(\pi_1, \pi_2)$, one of these corners of $\mathcal{P}$ will solve $L(\pi_1, \pi_2)$. Therefore, we can make a table of $f(\pi_1, \pi_2)$ as a function of the corners of $\mathcal{P}$.

| $(x_1, x_2)$ | $f(\pi_1, \pi_2)$ | $h$ |
|---|---|---|
| (0, 0) | $\pi_1 + 2\pi_2$ | $\binom{1}{2}$ |
| (1, 0) | $1 - 2\pi_1 + \pi_2$ | $\binom{-2}{1}$ |
| (0, 1) | $1 + 2\pi_1 - \pi_2$ | $\binom{2}{-1}$ |
| (1, 1) | $2 - \pi_1 - 2\pi_2$ | $\binom{-1}{-2}$ |

For each corner of $\mathcal{P}$, the set of $(\pi_1, \pi_2)$ for which that corner solves the $L(\pi_1, \pi_2)$ is the solution set of a finite number of linear inequalities. The following graph is a contour plot of $f$ for nonnegative $\pi$. The minimizing point of $f$ is at $(\pi_1, \pi_2) = (1/5, 2/5)$. The "cells" in the plot have been labeled with the corner of $\mathcal{P}$ associated with the solution of $L(\pi_1, \pi_2)$. Also, the table contains the gradient (hence, subgradient) $h$ of $f$ on the interior of each of the cells. Note that $z = 1$ $[(x_1, x_2) = (1/2, 1/2)$ is an optimal solution of $P]$ and that $f(1/5, 2/5) = 1$.

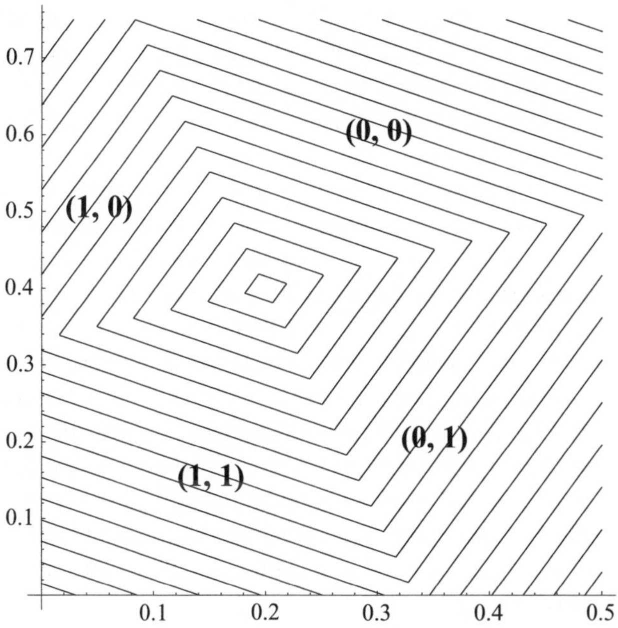


♠

**Problem (Subdifferential for the Lagrangian).** Suppose that the set of optimal solutions of $L(\pi)$ is bounded, and let $\widetilde{x}^1, \widetilde{x}^2, \ldots, \widetilde{x}^p$ be the extreme-point optima of $L(\pi)$. Let $h^k$ be defined by

$$h_i^k := b_i - \sum_{j=1}^{n} a_{ij}\widetilde{x}_j^k, \text{ for } i = 1, 2, \ldots, m.$$

Prove that the set of all subgradients of $f$ at $\pi$ is

$$\left\{ \sum_{k=1}^{p} \mu_k h^k \ : \ \sum_{k=1}^{p} \mu_k = 1, \ \mu_k \geq 0, \ k = 1, 2, \ldots, p \right\}.$$

**Exercise (Subdifferential for the Lagrangian).** For the Lagrangian Relaxation Example, calculate *all* subgradients of $f$ at *all* nonnegative $(\pi_1, \pi_2)$, and verify that 0 is a subgradient of $f$ at $\pi$ if and only if $(\pi_1, \pi_2) = (1/5, 2/5)$.

## 0.7 The Dual Simplex Method

An important tool in integer linear programming is a variant of the simplex method called the Dual Simplex Method. Its main use is to effectively resolve a linear program after an additional inequality is appended.

The Dual Simplex Method is initiated with a basic partition $\beta,\eta$ having the property that $\beta$ is dual feasible. Then, by a sequence of pairwise interchanges of elements between $\beta$ and $\eta$, the method attempts to transform $\beta$ into a primal-feasible basis, all the while maintaining dual feasibility. The index $\beta_i \in \beta$ is eligible to leave the basis if $x^*_{\beta_i} < 0$. If no such index exists, then $\beta$ is already primal feasible. Once the leaving index $\beta_i$ is selected, index $\eta_j \in \eta$ is eligible to enter the basis if $\Delta := -\overline{c}_{\eta_j}/\overline{a}_{i\eta_j}$ is minimized among all $\eta_j$ with $\overline{a}_{i\eta_j} < 0$. If no such index exists, then $P'$ is infeasible. If $\Delta > 0$, then the objective value of the primal solution decreases. As there are only a finite number of bases, this process must terminate in a finite number of iterations either with the conclusion that $P'$ is infeasible or with a basis that is both primal feasible and dual feasible. The only difficulty is that we may encounter "dual degeneracy" in which the minimum ratio $\Delta$ is zero. If this happens, then the objective value does not decrease, and we are not assured of finite termination. Fortunately, there is a way around this difficulty.

The Epsilon-Perturbed Dual Simplex Method is realized by application of the ordinary Dual Simplex Method to an algebraically perturbed program. Let

$$c'_j := c_j + \epsilon^j,$$

where $\epsilon$ is considered to be an arbitrarily small positive indeterminant. Consider applying the Dual Simplex Method with this new "objective row." After any sequence of pivots, components of the objective row will be polynomials in $\epsilon$. Ratios of real polynomials in $\epsilon$ form an ordered field – the ordering is achieved by considering $\epsilon$ to be an arbitrarily small positive real. Because the epsilons are confined to the objective row, every iterate will be feasible for the original problem. Moreover, the ordering of the field ensures that we are preserving dual feasibility even for $\epsilon = 0$. Therefore, even if $\epsilon$ were set to zero, we would be following the pivot rules of the Dual Simplex Method, maintaining dual feasibility, and the objective-function value would be nonincreasing.

The reduced cost $\overline{c}'_{\eta_j} = c'_j - c'_\beta \overline{A}_{\eta_j} = \overline{c}_{\eta_j} - \sum_{i=1}^m \overline{a}_{i\eta_j} \epsilon^{\beta^i} + \epsilon^{\eta_j}$ of nonbasic variable $x_{\eta_j}$ will never be identically zero because it always has an $\epsilon^{\eta_j}$ term. Therefore, the perturbed problem does not suffer from dual degeneracy, and the objective value of the basic solution $x^*$, which is $\sum_{j=1}^n c'_j x^*_j = \sum_{j=1}^n c_j x^*_j + \sum_{j=1}^n \epsilon^j x^*_j$, decreases at each iteration. Because there are only a finite number of bases, we have a finite version of the Dual Simplex Method.

## 0.8 Totally Unimodular Matrices, Graphs, and Digraphs

Some combinatorial-optimization problems can be solved with a straightforward application of linear-programming methods. In this section, the simplest phenomenon that leads to such fortuitous circumstances is described.

Let $A$ be the $m \times n$ real matrix with $a_{ij}$ in row $i$ and column $j$. The matrix $A$ is *totally unimodular* if every square nonsingular submatrix of $A$ has determinant $\pm 1$. Let $b$ be the $m \times 1$ vector with $b_i$ in position $i$. The following result can easily be established by use of Cramer's Rule.

**Theorem (Unimodularity implies integrality).** *If $A$ is totally unimodular and $b$ is integer valued, then every extreme point of $\mathcal{P}'$ is integer valued.*

*Proof.* Without loss of generality, we can assume that the rows of $A$ are linearly independent. Then every extreme point $x^*$ arises as the unique solution of

$$x^*_\eta = 0,$$
$$A_\beta x^*_\beta = b,$$

for some choice of basis $\beta$ and nonbasis $\eta$. By Cramer's Rule, the values of the basic variables are

$$x^*_{\beta_i} = \frac{\det(A^i_\beta)}{\det(A_\beta)}, \quad \text{for } i = 1, 2, \ldots, m,$$

where

$$A^i_\beta = \left[A_{\beta_1}, \ldots, A_{\beta_{i-1}}, b, A_{\beta_{i+1}}, \ldots, A_{\beta_m}\right].$$

Because $A^i_\beta$ is integer valued, we have that $\det(A^i_\beta)$ is an integer. Also, because $A$ is totally unimodular and $A_\beta$ is nonsingular, we have $\det(A_\beta) = \pm 1$. Therefore, $x^*_\beta$ is integer valued. ∎

A "near-converse" result is also easy to establish.

**Theorem (Integrality implies unimodularity).** *Let $A$ be an integer matrix. If the extreme points of $\{x \in \mathbf{R}^n : Ax \leq b,\ x \geq 0\}$ are integer valued, for all integer vectors $b$, then $A$ is totally unimodular.*

*Proof.* The hypothesis implies that the extreme points of

$$\mathcal{P}' := \{x \in \mathbf{R}^{n+m} : \widehat{A}x = b,\ x \geq 0\}$$

are integer valued for all integer vectors $b$, where $\widehat{A} := [A, I]$.

Let $B$ be an arbitrary invertible square submatrix of $\widehat{A}$. Let $\alpha'$ and $\beta'$ denote the row and column indices of $B$ in $\widehat{A}$. Using identity columns from $\widehat{A}$, we complete $B$ to an order-$m$ invertible matrix $\widehat{A}_\beta$ of $\widehat{A}$. Therefore, $\widehat{A}_\beta$ has the form

$$\begin{pmatrix} B & 0 \\ X & I \end{pmatrix}$$

Clearly we have $\det(\widehat{A}_\beta) = \pm\det(B)$.

Next, for each $i = 1, 2, \ldots, m$, we consider the right-hand side, $b := \Delta\widehat{A}_\beta \mathbf{e} + \mathbf{e}^i$, where $\Delta := \lceil \max_{k,j}(\widehat{A}_\beta^{-1})_{k,j} \rceil$. By the choice of $\Delta$, we have that the basic solution defined by the choice of basis $\beta$ is nonnegative: $x^*_\beta = \Delta e + \widehat{A}_\beta^{-1}\mathbf{e}^i$. Therefore, these basic solutions correspond to extreme points of $\mathcal{P}'$. By hypothesis, these are integer valued. Therefore, $x^*_\beta - \Delta\mathbf{e} = \widehat{A}_\beta^{-1}\mathbf{e}^i$ is also integer valued for $i = 1, 2, \ldots, m$. However, for each $i$, this is just the $i$th column of $\widehat{A}_\beta^{-1}$. Therefore, we conclude that $\widehat{A}_\beta^{-1}$ is integer valued.

Now, because $\widehat{A}_\beta$ and $\widehat{A}_\beta^{-1}$ are integer valued, each has an integer determinant. However, because the determinant of a matrix and its inverse are reciprocals of one another, they must both be $+1$ or $-1$. ∎

There are several operations that preserve total unimodularity. Obviously, if $A$ is totally unimodular, then $A^T$ and $[A, I]$ are also totally unimodular. Total unimodularity is also preserved under pivoting.

**Problem (Unimodularity and pivoting).** Prove that, if $A$ is totally unimodular, then any matrix obtained from $A$ by performing Gauss–Jordan pivots is also totally unimodular.

Totally unimodular matrices can also be joined together. For example, if $A^1$ and $A^2$ are totally unimodular, then

$$\begin{pmatrix} A^1 & 0 \\ 0 & A^2 \end{pmatrix}$$

is totally unimodular. The result of the following problem is more subtle.

**Problem (Unimodularity and connections).** The following $A^i$ are matrices, and the $b^i$ are row vectors. If

$$\begin{pmatrix} A^1 \\ b^1 \end{pmatrix} \quad \text{and} \quad \begin{pmatrix} b^2 \\ A^2 \end{pmatrix}$$

are totally unimodular, then

$$\begin{pmatrix} A^1 & 0 \\ b^1 & b^2 \\ 0 & A^2 \end{pmatrix}$$

is totally unimodular.

Some examples of totally unimodular matrices come from graphs and digraphs. To be precise, a *graph* or *digraph* $G$ consists of a finite set of vertices $V(G)$ and a finite collection of *edges* $E(G)$, the elements of which are pairs of elements of $V(G)$. For each edge of a digraph, one vertex is the *head* and the other is the *tail*. For each edge of a graph, both vertices are *heads*. Sometimes the vertices of an edge are referred to as its *endpoints*. If both endpoints of an edge are the same, then the edge is called a *loop*. The *vertex-edge incidence matrix* $A(G)$ of a graph or digraph is a $0, \pm 1$-valued matrix that has the rows indexed by $V(G)$ and the columns indexed by $E(G)$. If $v \in V(G)$ is a head (respectively, tail) of an edge $e$, then there is an additive contribution of $+1$ (respectively, $-1$) to $A_{ve}(G)$. Therefore, for a graph, every column of $A(G)$ has one $+1$ and one $-1$ entry – unless the column is indexed by a loop $e$ at $v$, in which case the column has no nonzeros, because $A_{ve}(G) = -1 + 1 = 0$. Similarly, for a digraph, every column of $A(G)$ has two $+1$ entries – unless the column is indexed by a loop $e$ at $v$, in which case the column has one nonzero value which is $A_{ve}(G) = +1 + 1 = +2$.

The most fundamental totally unimodular matrices derive from digraphs.

**Theorem (Unimodularity and digraphs).** *If $A(G)$ is the $0, \pm 1$-valued vertex-edge incidence matrix of a digraph $G$, then $A(G)$ is totally unimodular.*

*Proof.* We prove that each square submatrix $B$ of $A(G)$ has determinant $0, \pm 1$, by induction on the order $k$ of $B$. The result is clear for $k = 1$. There are three cases to consider: (1) If $B$ has a zero column, then $\det(B) = 0$; (2) if every column of $B$ has two nonzeros, then the rows of $B$ add up to the zero vector, so $\det(B) = 0$; (3) if some column of $B$ has exactly one nonzero, then we may expand the determinant along that column, and the result easily follows by use of the inductive hypothesis. ∎

A graph is *bipartite* if there is a partition of $V(G)$ into $V_1(G)$ and $V_2(G)$ (that is, $V(G) = V_1(G) \cup V_2(G)$, $V_1(G) \cap V_2(G) = \emptyset$, $E(G[V_1]) = E(G[V_2]) = \emptyset$), so that all edges have one vertex in $V_1(G)$ and one in $V_2(G)$. A *matching* of $G$ is a set of edges meeting each vertex of $G$ no more than once. A consequence of the previous result is the famous characterization of maximum-cardinality matchings in bipartite graphs.

**König's Theorem.** *The number of edges in a maximum-cardinality matching in a bipartite graph $G$ is equal to the minimum number of vertices needed to cover some endpoint of every edge of $G$.*

*Proof.* Let $G$ be a bipartite graph with vertex partition $V_1(G)$, $V_2(G)$. For a graph $G$ and $v \in V(G)$, we let $\delta_G(v)$ denote the set of edges of $G$ having exactly one endpoint at $v$. The maximum-cardinality matching problem is equivalent to solving the following program in *integer variables*:

$$\max \sum_{e \in E} x_e$$

subject to:

$$\begin{aligned}
\sum_{e \in \delta_G(v_1)} x_e \quad + s_{v_1} \quad\quad &= 1, && \forall\, v_1 \in V_1(G);\\
\sum_{e \in \delta_G(v_2)} x_e \quad\quad + s_{v_2} &= 1, && \forall\, v_2 \in V_2(G);\\
x_e &\geq 0, && \forall\, e \in E(G);\\
s_{v_1} &\geq 0, && \forall\, v_1 \in V_1(G);\\
s_{v_2} &\geq 0, && \forall\, v_2 \in V_2(G).
\end{aligned}$$

(We note that feasibility implies that the variables are bounded by 1.) The constraint matrix is totally unimodular (to see this, note that scaling rows or columns of a matrix by $-1$ preserves total unimodularity and that $A$ is totally unimodular if and only if $[A, I]$ is; we can scale the $V_1(G)$ rows by $-1$ and then scale the columns corresponding to $s_{v_1}$ variables by $-1$ to get a matrix that is of the form $[A, I]$, where $A$ is the vertex-edge incidence matrix of a digraph). Therefore, the optimal value of the program is the same as that of its linear-programming relaxation. Let $x$ be an optimal solution to this integer program. $S(x)$ is a maximum-cardinality matching of $G$. $|S(x)|$ is equal to the optimal objective value of the dual linear program

$$\min \sum_{v_1 \in V_1(G)} y_{v_1} + \sum_{v_2 \in V_2(G)} y_{v_2}$$

subject to:

$$\begin{array}{lll} y_{v_1} \quad + \quad y_{v_2} \geq 1, & & \forall\, e = \{v_1, v_2\} \in E(G); \\ y_{v_1} \geq 0, & & \forall\, v_1 \in V_1(G); \\ \qquad\qquad y_{v_2} \geq 0, & & \forall\, v_2 \in V_2(G). \end{array}$$

At optimality, no variable will have value greater than 1. Total unimodularity implies that there is an optimal solution that is integer valued. If $y$ is such a $0/1$-valued solution, then $S(y)$ meets every edge of $G$, and $|S(y)| = |S(x)|$.

■

**Hall's Theorem.** *Let $G$ be a bipartite graph with vertex partition $V_1(G)$, $V_2(G)$. The graph $G$ has a matching that meets all of $V_1(G)$ if and only if $|N(W)| \geq |W|$ for all $W \subset V_1(G)$.*

*Proof 1 (Hall's Theorem).* Clearly the condition $|N(W)| \geq |W|$, for all $W \subset V_1(G)$, is necessary for $G$ to contain a matching meeting all of $V_1(G)$. Therefore, suppose that $G$ has no matching that meets all of $V_1(G)$. Then the optimal objective value of the linear programs of the proof of König's Theorem is less than $|V_1(G)|$. As in that proof, we can select an optimal solution $y$ to the dual that is $0/1$-valued. That is, $y$ is a $0/1$-valued solution to

$$\sum_{v_1 \in V_1(G)} y_{v_1} + \sum_{v_2 \in V_2(G)} y_{v_2} < |V_1(G)|;$$

$$y_{v_1} + y_{v_2} \geq 1, \quad \forall\, e = (v_1, v_2) \in E(G).$$

Now, by defining $\widetilde{y} \in \{0, 1\}^{V(G)}$ by

$$\widetilde{y}_v := \begin{cases} 1 - y_v, & \text{for } v \in V_1(G) \\ y_v, & \text{for } v \in V_2(G) \end{cases},$$

we have a 0/1-valued solution $\widetilde{y}$ to

$$\sum_{v_2 \in V_2(G)} \widetilde{y}_{v_2} < \sum_{v_1 \in V_1(G)} \widetilde{y}_{v_1};$$
$$\widetilde{y}_{v_2} \geq \widetilde{y}_{v_1}, \quad \forall\, e = \{v_1, v_2\} \in E(G).$$

Let $W := \{v_1 \in V_1(G) : \widetilde{y}_{v_1} = 1\}$. The constraints clearly imply that $N(W) \subset \{v_2 \in V_2(G) : \widetilde{y}_{v_2} = 1\}$, and then that $|N(W)| < |W|$. ■

Hall's Theorem can also be proven directly from König's Theorem without a direct appeal to linear programming and total unimodularity.

*Proof 2 (Hall's Theorem).* Again, suppose that $G$ has no matching that meets all of $V_1(G)$. Let $S \subset V(G)$ be a minimum cover of $E(G)$. By König's Theorem, we have

$$|V_1(G)| > |S| = |S \cap V_1(G)| + |S \cap V_2(G)|.$$

Therefore,

$$|V_1(G) \setminus S| = |V_1(G)| - |S \cap V_1(G)| > |S \cap V_2(G)|.$$

Now, using the fact that $S$ covers $E(G)$, we see that $N(V_1(G) \setminus S) \subset S \cap V_2(G)$ (in fact, we have equality, as $S$ is a minimum-cardinality cover). Therefore, we just let $W := V_1(G) \setminus S$. ■

A 0, 1-valued matrix $A$ is a *consecutive-ones matrix* if the rows can be ordered so that the ones in each column appear consecutively.

**Theorem (Unimodularity and consecutive ones).** *If $A$ is a consecutive-ones matrix, then $A$ is totally unimodular.*

*Proof.* Let $B$ be a square submatrix of $A$. We may assume that the rows of $B$ are ordered so that the ones appear consecutively in each column. There are two cases to consider: (1) If the first row of $B$ is all zero, then $\det(B) = 0$; (2) if there is a one somewhere in the first row, consider the column $j$ of $B$ that has the least number, say $k$, of ones, among all columns with a one in the first row. Subtract row 1 from rows $i$ satisfying $2 \leq i \leq k$, and call the resulting matrix $B'$. Clearly $\det(B') = \det(B)$. By determinant expansion, we see that $\det(B') = \det(\widehat{B})$, where $\widehat{B}$ is the submatrix of $B'$ we obtain by deleting column $j$ and the first row. Now $\widehat{B}$ is a consecutive-ones matrix and its order is less than $B$, so the result follows by induction on the order of $B$. ■

**Example (Workforce planning).** Consecutive-ones matrices naturally arise in workforce planning. Suppose that we are planning for time periods $i = 1, 2, \ldots, m$. In time period $i$, we require that at least $d_i$ workers be assigned for work. We assume that workers can be hired for shifts of *consecutive* time periods and that the cost $c_j$ of staffing shift $j$ with each worker depends on only the shift. The number $n$ of shifts is at most $\binom{m+1}{2}$ – probably much less because an allowable shift may have restrictions (e.g., a maximum duration). The goal is to determine the number of workers $x_j$ to assign to shift $j$, for $j = 1, 2, \ldots, n$, so as to minimize total cost. We can formulate the problem as the integer linear program

$$\min \sum_{j=1}^{n} c_j x_j$$

subject to:

$$\sum_{j=1}^{n} a_{ij} x_j \geq d_i, \quad \text{for } i = 1, 2, \ldots, m;$$

$$x_j \geq 0 \text{ integer}, \quad \text{for } j = 1, 2, \ldots, n.$$

It is easy to see that the $m \times n$ matrix $A$ is a consecutive-ones matrix. Because such matrices are totally unimodular, we can solve the workforce-planning problem as a linear program. ♠

## 0.9 Further Study

There are several very important topics in linear programming that we have not even touched on. A course devoted to linear programming could certainly study the following topics:

1. Implementation issues connected with the Simplex Methods; in particular, basis factorization and updating, practical pivot rules (e.g., steepest-edge and devex), scaling, preprocessing, and efficient treatment of upper bounds on the variables.
2. Large-scale linear programming (e.g., column generation, Dantzig–Wolfe decomposition, and approximate methods).
3. Ellipsoid algorithms for linear programming; in particular, the theory and its consequences for combinatorial optimization.
4. Interior-point algorithms for linear programming; in particular, the theory as well as practical issues associated with its use.
5. The abstract combinatorial viewpoint of oriented matroids.

A terrific starting point for study of these areas is the survey paper by Todd (2002). The book by Grötschel, Lovász, and Schrijver (1988) is a great resource for topic 3. The monograph by Björner, Las Vergnas, Sturmfels, White and Ziegler (1999) is the definitive starting point for studying topic 5.

Regarding further study concerning the combinatorics of polytopes, an excellent resource is the book by Ziegler (1994).

The study of total unimodularity only begins to touch on the beautiful interplay between combinatories and integer linear programming. A much more thorough study is the excellent monograph by Cornuéjols (2001).

# 1

## Matroids and the Greedy Algorithm

Matroids are objects that generalize certain combinatorial aspects of linear dependence of finite sets of points in a vector space. A graph can be encoded by means of its 0/1-valued vertex-edge incidence matrix. It turns out that, when this matrix is viewed over **GF(2)**, each linearly independent set of columns corresponds to a forest in the underlying graph, and vice versa. Therefore, a fortiori, matroids generalize aspects of graphs. From this viewpoint, Hassler Whitney founded the subject of matroid theory in 1935.

In a natural sense, matroids turn out to yield the precise structure for which the most naïve "greedy" algorithm finds an optimal solution to combinatorial-optimization problems for all *weight* functions. Therefore, matroid theory is a natural starting point for studying combinatorial-optimization methods. Furthermore, matroids have algorithmic value well beyond the study of greedy algorithms (see, for example, Chapter 3).

In addition to the algorithmic importance of matroids, we also use matroids as a starting point for exploring the power of polytopes and linear-programming duality in combinatorial optimization.

### 1.1 Independence Axioms and Examples of Matroids

A *matroid* $M$ is a finite set $E(M)$ together with a subset $\mathcal{I}(M)$ of $2^{E(M)}$ that satisfies the following properties:

**Independence Axioms**

I1. $\emptyset \in \mathcal{I}(M)$.
I2. $X \subset Y \in \mathcal{I}(M) \Longrightarrow X \in \mathcal{I}(M)$.
I3. $X \in \mathcal{I}(M), Y \in \mathcal{I}(M), |Y| > |X| \Longrightarrow \exists\ e \in Y \setminus X$ such that $X + e \in \mathcal{I}(M)$.

The set $\mathcal{I}(M)$ is called the set of *independent sets* of $M$. The set $E(M)$ is called the *ground set* of $M$. Property I3 is called the *exchange axiom.*

What follows are some examples that we will revisit as we proceed.

**Example (Linear matroid).** Let $A$ be a matrix over a field $\mathbf{F}$, with columns indexed by the finite set $E(A)$. Let $E(M) := E(A)$, and let $\mathcal{I}(M)$ be the set of $X \subset E(M)$ such that the columns of $A_X$ are linearly independent. In this case, we say that $M$ is the *linear matroid* of $A$ and that $A$ is a *representation* of $M$ over $\mathbf{F}$. It is very easy to see that properties I1 and I2 hold. To see how I3 holds, suppose that $X + e \notin \mathcal{I}(M)$ for every $e \in Y \setminus X$. Then the columns of $A_Y$ are in c.s.$(A_X)$ (the *column space* or *linear span* of $A_X$). Hence, c.s.$(A_Y)$ is a subset of c.s.$(A_X)$. Therefore, the dimension of c.s.$(A_Y)$ is no more than that of c.s.$(A_X)$. Therefore, we have $|Y| \leq |X|$. ♠

Let $G$ be a graph with vertex set $V(G)$ and edge set $E(G)$. We denote the numbers of connected components of $G$ (counting isolated vertices as components) by $\kappa(G)$. For $F \subset E(G)$, let $G.F$ ($G$ *restricted to* $F$) denote the graph with $V(G.F) := V(G)$ and $E(G.F) := F$. A set of edges $F$ of graph $G$ is a *forest* if it contains no cycle.

**Lemma (Forest components).** *Let $F$ be a forest of a graph $G$. Then $|F| = |V(G)| - \kappa(G.F)$.*

*Proof.* By induction of $|F|$. Clearly true for $|F| = 0$. For the inductive step, we just observe that, for $e \in F$, $\kappa(G.(F - e)) = \kappa(G.F) - 1$. ∎

**Example (Graphic matroid).** Let $G$ be a graph. Let $E(M) := E(G)$, and let $\mathcal{I}(M)$ be the set of forests of $G$. In this case, we say that $M$ is the *graphic matroid* of $G$. It is easy to see that I1 and I2 hold. To see how I3 holds, suppose that $X$ and $Y$ are forests such that $X + e$ is not a forest for every $e \in Y \setminus X$. Then every edge in $Y \setminus X$ would have both ends in the same component of $G.X$. Hence, $\kappa(G.Y) \geq \kappa(G.X)$. Therefore, by the Lemma (Forest components), we have $|Y| \leq |X|$. ♠

**Example (Uniform matroid).** Let $E(M)$ be a finite set, and let $r$ be an integer satisfying $0 \leq r \leq |E(M)|$. Let $\mathcal{I}(M) := \{X \subset E(M) \ : \ |X| \leq r\}$. In this case, we say that $M$ is a *uniform matroid.* ♠

**Example (Direct sum).** Let $M_1$ and $M_2$ be matroids with $E(M_1) \cap E(M_2) = \emptyset$. Define $M$ by $E(M) := E(M_1) \cup E(M_2)$, and $\mathcal{I}(M) := \{X_1 \cup X_2 \ : \ X_1 \in \mathcal{I}(M_1),\ X_2 \in \mathcal{I}(M_2)\}$. Then matroid $M$ is the *direct sum* of $M_1$ and $M_2$. ♠

A system that respects I1 and I2 but not necessarily I3 is called an *independence system*. As the following example indicates, not every independence system is a matroid.

**Example (Vertex packing on a star).** Let $G$ be a simple undirected graph. Define $M$ by $E(M) := V(G)$, and let $\mathcal{I}(M)$ be the set of "vertex packings" of $G$ – a *vertex packing* of $G$ is just a set of vertices $X$ with no edges of $G$ between elements of $X$. Clearly $M$ is an independence system. To see that $M$ need not be a matroid consider the *$n$-star graph*:

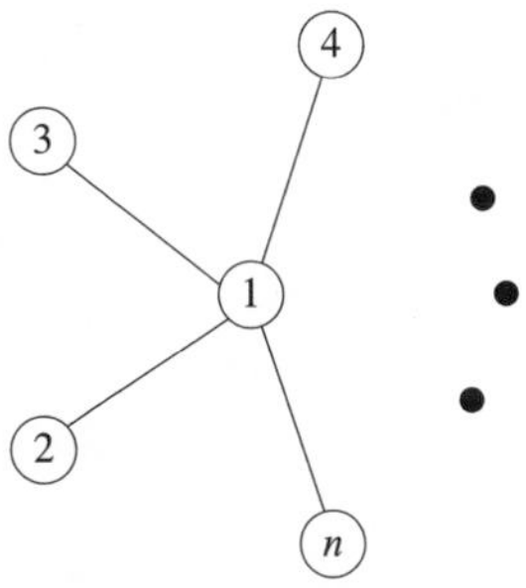


with $n \geq 3$. The pair $X = \{1\}$, $Y = \{2, 3, \ldots, n\}$ violates I3. ♠

## 1.2 Circuit Properties

For any independence system, the elements of $2^{E(M)} \setminus \mathcal{I}(M)$ are called the *dependent sets* of $M$. We distinguish the dependent sets whose proper subsets are in $\mathcal{I}(M)$. We call these subsets the *circuits* of $M$, and we write the set of circuits of $M$ as

$$\mathcal{C}(M) := \{X \subset E(M) \; : \; X \notin \mathcal{I}(M), \; X - e \in \mathcal{I}(M), \quad \forall\, e \in X\}.$$

For example, if $M$ is the graphic matroid of a graph $G$, then the circuits of $M$ are the cycles of $G$. Single-element circuits of a matroid are called *loops*; if $M$ is the graphic matroid of a graph $G$, then the set of loops of $M$ is precisely the set of loops of $G$.

**Problem [Graphic $\Longrightarrow$ linear over GF(2)].** Show that if $A(G)$ is the vertex-edge incidence matrix of $G$, then the matroid represented by $A(G)$, with numbers of $A(G)$ interpreted in **GF(2)**, is precisely the graphic matroid of $G$.

If $M$ is a matroid, then $\mathcal{C}(M)$ obeys the following properties:

**Circuit Properties**

C1. $\emptyset \notin \mathcal{C}(M)$.
C2. $X \in \mathcal{C}(M)$, $Y \in \mathcal{C}(M)$, $X \subset Y \Longrightarrow X = Y$.
C3. $X \in \mathcal{C}(M)$, $Y \in \mathcal{C}(M)$, $X \neq Y$, $e \in X \cap Y \Longrightarrow \exists\ Z \subset (X \cup Y) - e$ such that $Z \in \mathcal{C}(M)$.

Properties C1 and C2 follow from I1 and I2 and the definition of $\mathcal{C}(M)$.

**Theorem (Circuit elimination).** *If $M$ is a matroid, then $\mathcal{C}(M)$ satisfies C3.*

*Proof.* Suppose that $X$, $Y$, $e$ satisfy the hypotheses of C3 but that $(X \cup Y) - e$ contains no element of $\mathcal{C}(M)$. By C2, $Y \setminus X \neq \emptyset$, so choose some $f \in Y \setminus X$. By the definition of $\mathcal{C}(M)$, $Y - f \in \mathcal{I}(M)$.

Let $W$ be a subset of $X \cup Y$ that is maximal among all sets in $\mathcal{I}(M)$ that contain $Y - f$. Clearly $f \notin W$. Choose some $g \in X \setminus W$ [the set $X \setminus W$ is nonempty because $X$ is a circuit and $W \in \mathcal{I}(M)$]. Clearly $f$ and $g$ are distinct because $f \in Y \setminus X$. In the following figure $W$ is indicated by the shaded region.

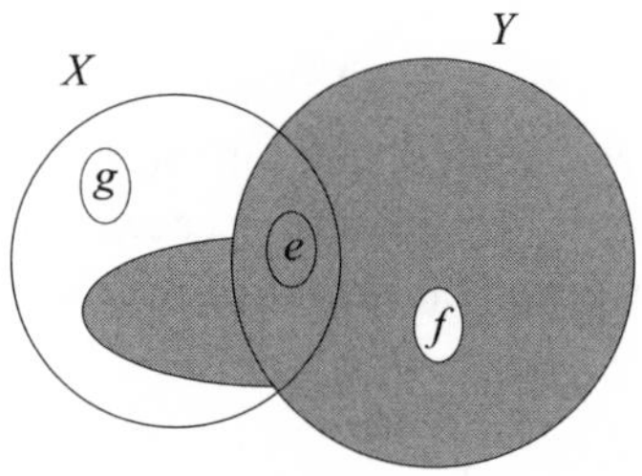


Hence,

$$|W| \leq |(X \cup Y) \setminus \{f, g\}| = |X \cup Y| - 2 < |(X \cup Y) - e|.$$

Now, applying I3 to $W$ and $(X \cup Y) - e$, we see that there is an element $h \in ((X \cup Y) - e) \setminus W$, such that $W + h \in \mathcal{I}(M)$. This contradicts the maximality of $W$. ∎

**Problem (Linear circuit elimination).** Give a direct proof of C3 for linear matroids.

**Problem (Graphic circuit elimination).** Give a direct proof of C3 for graphic matroids.

Property C3 is called the *circuit-elimination* property. A system satisfying properties C1 and C2 but not necessarily C3 is called a *clutter*.

**Example [Vertex packing on a star, continued (see p. 51)].** $X := \{1, i\}$ and $Y := \{1, j\}$ are distinct circuits for $1 \neq i \neq j \neq 1$, but $\{i, j\}$ contains no circuit. ♠

It should be clear that $\mathcal{C}(M)$ completely determines $\mathcal{I}(M)$ for any independence system. That is, given $E(M)$ and $\mathcal{C}(M)$ satisfying C1 and C2, there is precisely one choice of $\mathcal{I}(M)$ that has circuit set $\mathcal{C}(M)$ that will satisfy I1 and I2. That choice is

$$\mathcal{I}(M) := \{X \subset E(M) \; : \nexists \, Y \subset X, \; Y \in \mathcal{C}(M)\}.$$

**Problem (Unique-circuit property).** Let $M$ be a matroid. Prove that if $X \in \mathcal{I}(M)$ and $X + e \notin \mathcal{I}(M)$, then $X + e$ contains a unique circuit of $M$. Give an example to show how this need not hold for a general independence system.

**Problem (Linear unique circuit).** Give a direct proof of the unique-circuit property for linear matroids.

**Problem (Graphic unique circuit).** Give a direct proof of the unique-circuit property for graphic matroids.

## 1.3 Representations

The *Fano matroid* is the matroid represented over **GF(2)** by the matrix

$$F = \begin{array}{c} \begin{matrix} 1 & 2 & 3 & 4 & 5 & 6 & 7 \end{matrix} \\ \begin{pmatrix} 1 & 0 & 0 & 0 & 1 & 1 & 1 \\ 0 & 1 & 0 & 1 & 0 & 1 & 1 \\ 0 & 0 & 1 & 1 & 1 & 0 & 1 \end{pmatrix} \end{array}.$$

**Exercise [Linear over GF(2) $\not\Longrightarrow$ graphic].** Prove that the Fano matroid is not graphic.

A linear matroid may have many representations. A *minimal representation* of $M$ is a representation having linearly independent rows. If $A$ and $A'$ are $r \times n$ matrices over the same field, having full row rank, and there is a nonsingular matrix $B$ and a nonsingular diagonal matrix $D$ such that $A' = BAD$, then $A$ and $A'$ are *projectively equivalent*. It is easy to see that projective equivalence is an equivalence relation. If $A$ and $A'$ are projectively equivalent then they represent the same matroid; however, the converse is not generally true.

**Proposition (Fano representation).** *The Fano matroid is representable over a field if and only if the field has characteristic 2. Moreover, $F$ is the only minimal representation of the Fano matroid over every characteristic-2 field, up to projective equivalence.*

*Proof.* If the Fano matroid can be represented over a field $\mathbf{F}$, then it has a minimal representation over $\mathbf{F}$ of the form

$$A = \begin{array}{c} \begin{array}{ccccccc} 1 & 2 & 3 & 4 & 5 & 6 & 7 \end{array} \\ \begin{pmatrix} a_{11} & a_{12} & a_{13} & a_{14} & a_{15} & a_{16} & a_{17} \\ a_{21} & a_{22} & a_{23} & a_{24} & a_{25} & a_{26} & a_{27} \\ a_{31} & a_{32} & a_{33} & a_{34} & a_{35} & a_{36} & a_{37} \end{pmatrix} \end{array}.$$

The first three columns of $A$ must be linearly independent, so, by using elementary row operations, we can bring $A$ into the form

$$A' = \begin{array}{c} \begin{array}{ccccccc} 1 & 2 & 3 & 4 & 5 & 6 & 7 \end{array} \\ \begin{pmatrix} 1 & 0 & 0 & a'_{14} & a'_{15} & a'_{16} & a'_{17} \\ 0 & 1 & 0 & a'_{24} & a'_{25} & a'_{26} & a'_{27} \\ 0 & 0 & 1 & a'_{34} & a'_{35} & a'_{36} & a'_{37} \end{pmatrix} \end{array}.$$

We have $a'_{14} = 0$, $a'_{24} \neq 0$, and $a'_{34} \neq 0$, as $\{2, 3, 4\}$ is a circuit. Similarly, $a'_{15} \neq 0$, $a'_{25} = 0$, $a'_{35} \neq 0$, and $a'_{16} \neq 0$, $a'_{26} \neq 0$, $a'_{36} = 0$. Finally, $a'_{17} \neq 0$, $a'_{27} \neq 0$, and $a'_{37} \neq 0$, as $\{1, 2, 3, 7\}$ is a circuit.

Therefore, any minimal representation of the Fano matroid over a field **F**, up to multiplication on the left by an invertible matrix, is of the form

$$\begin{array}{ccccccc} 1 & 2 & 3 & 4 & 5 & 6 & 7 \end{array}$$
$$\begin{pmatrix} 1 & 0 & 0 & 0 & a & b & c \\ 0 & 1 & 0 & d & 0 & e & f \\ 0 & 0 & 1 & g & h & 0 & i \end{pmatrix},$$

with the letters being nonzeros in the field **F**. We can bring the matrix into the form

$$\begin{array}{ccccccc} 1 & 2 & 3 & 4 & 5 & 6 & 7 \end{array}$$
$$\begin{pmatrix} 1 & 0 & 0 & 0 & 1 & 1 & 1 \\ 0 & 1 & 0 & 1 & 0 & q & 1 \\ 0 & 0 & 1 & r & s & 0 & 1 \end{pmatrix},$$

with the letters being nonzeros, by nonzero row and column scaling (multiply row 1 by $c^{-1}$, row 2 by $f^{-1}$, row 3 by $i^{-1}$, column 4 by $d^{-1}f$, column 5 by $a^{-1}c$, column 6 by $b^{-1}c$, column 1 by $c$, column 2 by $f$, and column 3 by $i$).

Now, columns 1, 4, and 7 should be dependent; calculating the determinant and setting it to 0, we get $r = 1$. Similarly, the required dependence of columns 2, 5, and 7 implies $s = 1$, and the dependence of columns 3, 6, and 7 implies $q = 1$. Therefore, over any field **F**, $F$ is the only minimal representation of the Fano matroid, up to projective equivalence.

Finally, columns 4, 5, and 6 should be dependent, so we get $1 + 1 = 0$. Therefore, the field must have characteristic 2. ■

The *non-Fano matroid* arises when the **GF(2)** representation of the Fano matroid is used but the numbers are considered as rational. The representation $F$, viewed over **Q**, is projectively equivalent to the rational matrix

$$\begin{array}{ccccccc} 1 & 2 & 3 & 4 & 5 & 6 & 7 \end{array}$$
$$F_- = \begin{pmatrix} 1 & 0 & 0 & 0 & 1/2 & 1/2 & 1/3 \\ 0 & 1 & 0 & 1/2 & 0 & 1/2 & 1/3 \\ 1 & 1 & 1 & 1 & 1 & 1 & 1 \end{pmatrix}.$$

Let $F'_-$ be the matrix that we obtain by deleting the last row (of all 1's) of $F_-$. The *linear* dependencies among the columns of $F_-$ are the same as the *affine* dependencies among the columns of the matrix $F'_-$. We can plot the columns of $F'_-$ as points in the Euclidean plane and then visualize the independent sets of the non-Fano matroid as the sets of points that are affinely independent (in the plane, this means pairs of points that are not coincident and triples of points that do not lie on a straight line):

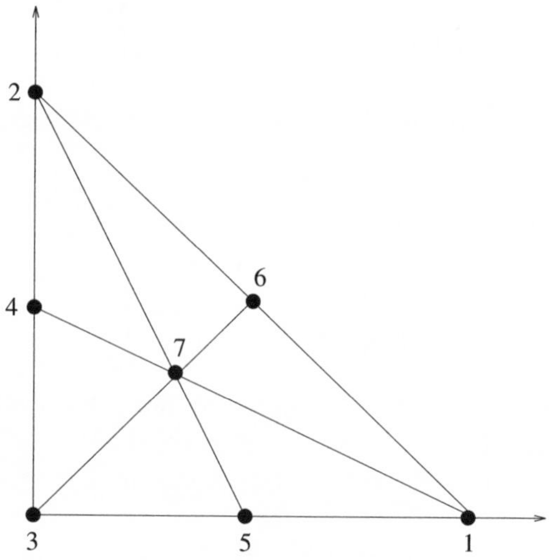


**Exercise (Nonrepresentable matroids).** First prove that the non-Fano matroid is representable over a field if and only if the characteristic of the field is not 2, and then prove that there are matroids representable over no field by taking the direct sum of the Fano matroid and the non-Fano matroid.

## 1.4 The Greedy Algorithm

Associated with any independence system $M$ is its *rank function* $r_M : 2^{E(M)} \mapsto \mathbf{R}$, defined by

$$r_M(X) := \max\{|Y| \ : \ Y \subset X, \ Y \in \mathcal{I}(M)\}.$$

We call $r_M(E(M))$ the *rank* of $M$. A set $S \subset E(M)$ such that $S \in \mathcal{I}(M)$ and $|S| = r_M(E(M))$ is a *base* of $M$. We write $\mathcal{B}(M)$ for the set of bases of $M$. It is a simple matter to find a base of the independence system $M$ when $M$ is a matroid, provided that we can easily recognize when a set is in $\mathcal{I}(M)$. We simply use a "greedy" algorithm:

**Cardinality Greedy Algorithm**

1. $S := \emptyset$. $U := E(M)$.
2. While ($U \neq \emptyset$)
   i. choose any $e \in U$; $U := U - e$;
   ii. if $S + e \in \mathcal{I}(M)$, then $S := S + e$.

Throughout execution of the algorithm, $S \subset E(M)$ and $S \in \mathcal{I}(M)$. At termination, $|S| = r_M(E(M))$ (convince yourself of this by using I2 and I3).

The algorithm need not find a base of $M$, if $M$ is a general independence system.

**Example [Vertex packing on a star, continued (see pp. 51, 53)].** If 1 is chosen as the first element to put in $S$, then no other element can be added, but the only base of $M$ is $\{2, 3, \ldots, n\}$. ♠

With respect to a matroid $M$ and weight function $c$, we consider the problem of finding maximum-weight independent sets $S_k$ of cardinality $k$ for all $k$ satisfying $0 \leq k \leq r_M(E(M))$. This is an extension of the problem of determining the rank of $M$; in that case, $c(\{e\}) = 1$, $\forall\, e \in E(M)$, and we concern ourselves only with $k = r_M(E(M))$. A greedy algorithm for the present problem is as follows:

**(Weighted) Greedy Algorithm**

1. $S_0 := \emptyset$. $k := 1$. $U := E(M)$.
2. While ($U \neq \emptyset$)
   i. choose $s_k \in U$ of maximum weight; $U := U - s_k$;
   ii. if $S_{k-1} + s_k \in \mathcal{I}(M)$, then $S_k := S_{k-1} + s_k$ and $k := k + 1$.

Next we demonstrate that each time an $S_k$ is assigned, $S_k$ is a maximum-weight independent set of cardinality $k$.

**Theorem (Greedy optimality for matroids).** *The Greedy Algorithm finds maximum-weight independent sets of cardinality $k$ for every $k$ satisfying $1 \leq k \leq r_M(E(M))$.*

*Proof.* The proof is by contradiction. Note that $S_k = \{s_1, s_2, \ldots, s_k\}$ for $1 \leq k \leq r_M(E(M))$. Hence, $c(s_1) \geq c(s_2) \geq \cdots \geq c(s_k)$. Let $T_k = \{t_1^k, t_2^k, \ldots, t_k^k\}$ be any maximum-weight independent set of cardinality $k$, with the elements numbered so that $c(t_1^k) \geq c(t_2^k) \geq \cdots \geq c(t_k^k)$. Suppose that $c(T_k) > c(S_k)$; then there exists $p$, $1 \leq p \leq k$, such that $c(t_p^k) > c(s_p)$. Now, consider the sets

$$\left\{t_1^k, t_2^k, \ldots, t_{p-1}^k, t_p^k\right\},$$
$$\{s_1, s_2, \ldots, s_{p-1}\}.$$

Property I3 implies that there is some $i$, $1 \leq i \leq p$, such that

$$t_i^k \notin \{s_1, s_2, \ldots, s_{p-1}\},$$
$$\{s_1, s_2, \ldots, s_{p-1}\} + t_i^k \in \mathcal{I}(M).$$

Now $c(t_i^k) \geq c(t_{i+1}^k) \geq \cdots \geq c(t_p^k) > c(s_p)$; therefore, $t_i^k$ should have been chosen in preference to $s_p$ by the Greedy Algorithm. ■

**Exercise (Maximum-weight spanning tree).** Use the Greedy Algorithm, with respect to the graphic matroid of the following edge-weighted graph, to find a maximum-weight spanning tree.

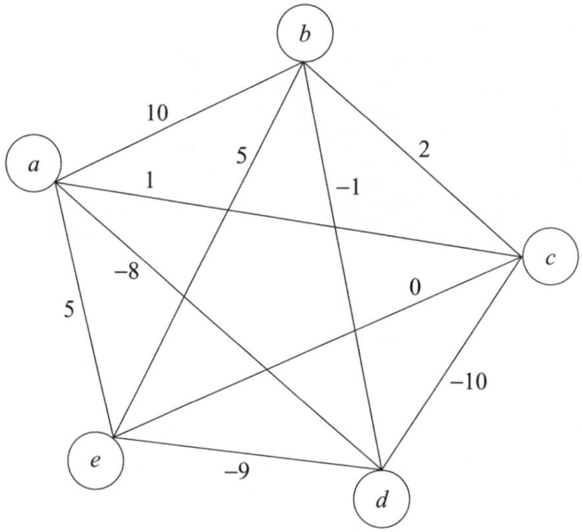

The Greedy Algorithm can be used to find a maximum-weight independent set (with no restriction on the cardinality) by stopping once all positive-weight elements have been considered in Step 2.i.

**Problem (Scheduling).** Jobs labeled $1, 2, \ldots, n$ are to be processed by a single machine. All jobs require the same processing time. Each job $j$ has a deadline $d_j$ and a profit $c_j$, which will be earned if the job is completed by its deadline. The problem is to find the ordering of the jobs that maximizes total profit. First, prove that if a subset of the jobs can be completed on time, then they will be completed on time if they are ordered by deadline. Next, let $E(M) := \{1, 2, \ldots, n\}$, and

$$\mathcal{I}(M) := \{J \subset E(M) \; : \; \text{the jobs in } J \text{ are completed on time}\}.$$

Prove that $M$ is a matroid by verifying that I1–I3 hold for $\mathcal{I}(M)$, and describe a method for finding an optimal order for processing the jobs.

**Exercise (Scheduling).** Solve the scheduling problem with the following data. The machine is available at 12:00 noon, and each job requires one hour of processing time.

| Job $j$ | $c_j$ | $d_j$ |
|---|---|---|
| 1 | 20 | 3:00 P.M. |
| 2 | 15 | 1:00 P.M. |
| 3 | 10 | 2:00 P.M. |
| 4 | 10 | 1:00 P.M. |
| 5 | 6 | 2:00 P.M. |
| 6 | 4 | 5:00 P.M. |
| 7 | 3 | 5:00 P.M. |
| 8 | 2 | 4:00 P.M. |
| 9 | 2 | 2:00 P.M. |
| 10 | 1 | 6:00 P.M. |

It is natural to wonder whether some class of independence systems, more general than matroids, might permit the Greedy Algorithm to always find maximum-weight independent sets of all cardinalities. The following result ends such speculation.

**Theorem (Greedy characterization of matroids).** *Let $M$ be an independence system. If the Greedy Algorithm produces maximum-weight independent sets of all cardinalities for every (nonnegative) weight function, then $M$ is a matroid.*

*Proof.* We must prove that $\mathcal{I}(M)$ satisfies I3. The proof is by contradiction. Choose $Y$ and $X$ so that I3 fails. We assign weights as follows:

$$c(e) := \begin{cases} 1+\epsilon, & \text{if } e \in X \\ 1, & \text{if } e \in Y \setminus X \\ 0, & \text{if } e \in E(M) \setminus (X \cup Y) \end{cases},$$

with $\epsilon > 0$ to be determined. Because $Y$ is in $\mathcal{I}(M)$, the Greedy Algorithm should find a maximum-weight independent set of cardinality $|Y|$. With just $|X|$ steps, the Greedy Algorithm chooses all of $X$, and then it completes $X$ to an independent set $X'$ of cardinality $|Y|$ by using 0-weight elements, for a total weight of $|X|(1+\epsilon)$. Now we just take care to choose $\epsilon < 1/|E(M)|$, so that $c(X') < c(Y)$. This is a contradiction. ∎

**Problem (Swapping Algorithm)**

<u>**Swapping Algorithm**</u>

1. Choose any $S \in \mathcal{I}(M)$, such that $|S| = k$.
2. While ($\exists\, S' \in \mathcal{I}(M)$ with $|S'| = k$, $|S \Delta S'| = 2$ and $c(S') > c(S)$): Let $S := S'$.

Prove that if $M$ is a matroid, then the Swapping Algorithm finds a maximum-weight independent set of cardinality $k$.

**Exercise [Maximum-weight spanning tree, continued (see p. 58)].** Apply the Swapping Algorithm to calculate a maximum-weight spanning tree for the edge-weighted graph of the Maximum-weight spanning tree Exercise.

## 1.5 Rank Properties

Let $E$ be a finite set, and let $M$ be a matroid with $E(M) = E$. If $r := r_M$, then $r$ satisfies the following useful properties:

**Rank Properties**

R1. $0 \leq r(X) \leq |X|$, and integer valued, $\forall\, X \subset E$.
R2. $X \subset Y \Longrightarrow r(X) \leq r(Y)$, $\forall\, X, Y \subset E$.
R3. $r(X \cup Y) + r(X \cap Y) \leq r(X) + r(Y)$, $\forall\, X, Y \subset E$.

Property R3 is called *submodularity*. The rank function of a general independence system $M$ need only satisfy R1 and R2 and the weaker property of *subadditivity*: $r_M(X \cup Y) \leq r_M(X) + r_M(Y)$.

**Example [Vertex packing on a star, continued (see pp. 51, 53, 57)].** For $X := \{1, i\}$ and $Y := \{1, j\}$, with $i \neq j$, we have $r_M(X) = 1$, $r_M(Y) = 1$, $r_M(X \cup Y) = 2$, and $r_M(X \cap Y) = 1$. ♠

**Problem (Cuts).** Let $G$ be a graph, let $E := V(G)$, let $c$ be a nonnegative-weight function on $E(G)$, and define $r(X) := \sum_{e \in \delta_G(X)} c(e)$, for $X \subset E$. Show that $r$ always satisfies R3, but need not satisfy R1 and R2 [even when $c(e) = 1$, for all $e \in E(G)$].

**Theorem (Submodularity of matroid rank function).** *If $M$ is a matroid, then $r_M$ satisfies R3.*

*Proof.* Let $J$ be a maximal independent subset of $X \cap Y$. Extend $J$ to $J_X$ ($J_Y$), a maximal independent subset of $X$ ($Y$, respectively). We have $r_M(X \cap Y) = |J| = |J_X \cap J_Y|$. If we can show that $r_M(X \cup Y) \leq |J_X \cup J_Y|$, then R3 follows, because $|J_X \cup J_Y| + |J_X \cap J_Y| = |J_X| + |J_Y|$. Extend $J_X$ to a maximal independent subset $K$ of $X \cup Y$.

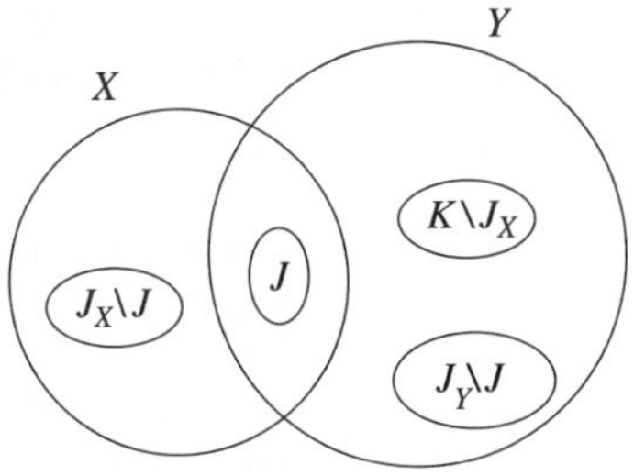

Suppose that $|K| > |J_X \cup J_Y|$. Because $J_X \setminus J$ is contained in both $K$ and $J_X \cup J_Y$, we have $|K \setminus (J_X \setminus J)| > |J_Y|$. Now, by the choice of $J_X$, we have that $K \setminus (J_X \setminus J)$ is an independent subset of $Y$. This contradicts the choice of $J_Y$. ■

Our next goal is to show that R1–R3 characterize the rank functions of matroids. That is, for every $E$ and $r$ satisfying R1–R3, there is a matroid $M$ with $E(M) = E$ and $r_{_M} = r$. First, we establish a useful lemma.

**Lemma (Closure).** *Suppose that $r : 2^E \mapsto \mathbf{R}$ satisfies R2 and R3. If $X$ and $Y$ are arbitrary subsets of $E$ with the property that $r(X + e) = r(X)$, $\forall\, e \in Y \setminus X$, then $r(X \cup Y) = r(X)$.*

*Proof.* The proof is by induction on $k = |Y \setminus X|$. For $k = 1$ there is nothing to show. For $k > 1$, choose $e \in Y \setminus X$.

$$\begin{aligned} 2r(X) &= r(X \cup ((Y \setminus X) - e)) + r(X + e) \quad \text{(by the inductive hypothesis)} \\ &\geq r(X \cup Y) + r(X) \quad \text{(by R3)} \\ &\geq 2r(X) \quad \text{(by R2).} \end{aligned}$$

Therefore, equality must hold throughout, and we conclude that $r(X \cup Y) = r(X)$. ■

**Theorem (Rank characterization of matroids).** *Let $E$ be a finite set, and suppose that $r : 2^E \mapsto \mathbf{R}$ satisfies R1–R3. Then*

$$\mathcal{I}(M) := \{Y \subset E(M) \ : |Y| = r(Y)\}.$$

*defines a matroid $M$ with $E(M) := E$ and $r_{_M} = r$.*

*Proof.* We show that the choice of $\mathcal{I}(M)$ in the statement of the theorem satisfies I1–I3, and then show that $r$ is indeed the rank function of $M$.

R1 implies that $r(\emptyset) = 0$; therefore, $\emptyset \in \mathcal{I}(M)$, and I1 holds for $\mathcal{I}(M)$.

Now, suppose that $X \subset Y \in \mathcal{I}(M)$. Therefore, $r(Y) = |Y|$. R3 implies that

$$r(X \cup (Y \setminus X)) + r(X \cap (Y \setminus X)) \leq r(X) + r(Y \setminus X),$$

which reduces to

$$r(Y) \leq r(X) + r(Y \setminus X).$$

Using the facts that $r(Y) = |Y|$, $r(Y \setminus X) \leq |Y \setminus X|$, and $r(X) \leq |X|$, we can conclude that $r(X) = |X|$. Therefore, $X \in \mathcal{I}(M)$, and I2 holds for $\mathcal{I}(M)$.

Next, choose arbitrary $X, Y \in \mathcal{I}(M)$, such that $|Y| > |X|$. We prove I3 by contradiction. If I3 fails, then $r(X + e) = r(X)$ for all $e \in Y \setminus X$. Applying the Closure Lemma, we have $r(X \cup Y) = r(X)$. However, $r(X) = |X|$ and $r(X \cup Y) \geq r(Y) = |Y|$ implies $|Y| \leq |X|$. Therefore, I3 holds for $\mathcal{I}(M)$.

We conclude that $M$ is a matroid on $E$. Because $M$ is a matroid, it has a well-defined rank function $r_M$ which satisfies

$$r_M(Y) = \max\{|X| \ : X \subset Y, \ |X| = r(X)\}.$$

R2 for $r$ implies that

$$\max\{|X| \ : X \subset Y, \ |X| = r(X)\} \leq r(Y).$$

Therefore, we need show only that $Y$ contains a set $X$ with $|X| = r(X) = r(Y)$. Let $X$ be a maximal independent subset of $Y$. Because $X + e \notin \mathcal{I}(M)$, $\forall\, e \in Y \setminus X$, we have $r(X + e) = r(X)$, $\forall\, e \in Y \setminus X$. By the Closure Lemma, we can conclude that $r(Y) = r(X) = |X|$, and we are done. ■

## 1.6 Duality

Every matroid $M$ has a natural *dual* $M^*$ with $E(M^*) := E(M)$ and

$$\mathcal{I}(M^*) := \{X \subset E(M) \ : \ E(M) \setminus X \text{ contains a base of } M\}.$$

**Theorem (Matroid duality).** *The dual of a matroid is a matroid.*

*Proof.* Clearly $M^*$ is an independence system. Therefore, it possesses a well-defined rank function $r_{M^*}$. First we demonstrate that

$$r_{M^*}(X) = |X| + r_M(E(M) \setminus X) - r_M(E(M)), \quad \forall\, X \subset E(M^*).$$

Let $Y$ be a subset of $X$ that is in $\mathcal{I}(M^*)$. By the definition of $\mathcal{I}(M^*)$, $E(M) \setminus Y$ contains a base $B$ of $M$. If $Y$ is a (setwise) maximal subset of $X$ that is in $\mathcal{I}(M^*)$, then $(X \setminus B) \setminus Y$ is empty (otherwise we could append such elements to $Y$ to get a larger set). Therefore, a maximal such $Y$ is of the form $X \setminus B$ for some base $B$ of $M$. Now, if $Y = X \setminus B$ is a maximum cardinality such set, then

$|X \cap B|$ must be a small as possible, because all bases of $M$ have the same cardinality.

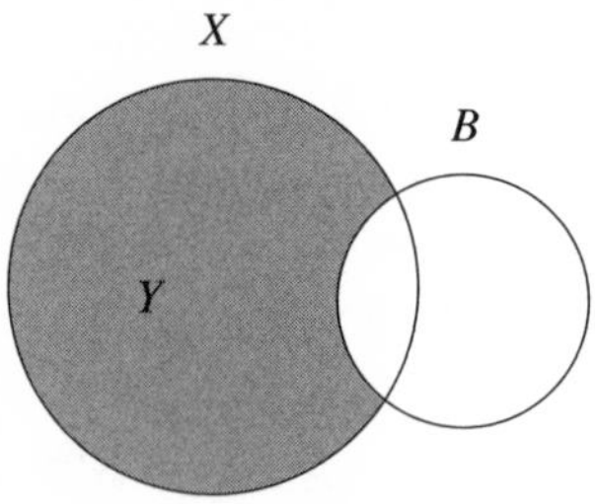


Therefore, for any $X \subset E(M)$, let $B_X$ be a base of $M$ with $|B_X \cap X|$ as small as possible. By the choice of $B_X$, we have $r_{M^*}(X) = |X \setminus B_X|$. Moreover, the choice of $B_X$ dictates that $|B_X \setminus X| = r_M(E(M) \setminus X)$. Therefore, we have

$$\begin{aligned} r_{M^*}(X) &= |X \setminus B_X| \\ &= |X| + |B_X \setminus X| - |B_X| \\ &= |X| + r_M(E(M) \setminus X) - r_M(E(M)). \end{aligned}$$

We leave verification that $r := r_{M^*}$ satisfies R1–R3 as a problem. ■

**Problem (Dual rank function).** Verify that $r := r_{M^*}$ satisfies R1–R3 when $M$ is a matroid.

It is clear from the specification of $\mathcal{I}(M^*)$ that the bases of $M^*$ are precisely the complements of the bases of $M$. That is, $\mathcal{B}(M^*) = \{E(M) \setminus B \ : \ B \in \mathcal{B}(M)\}$. Therefore, another algorithm for finding a maximum-weight base $B$ of $M$, with respect to the weight function $c$, is to use the Greedy Algorithm to select a *minimum*-weight base $B^*$ of $M^*$, and then let $B := E(M) \setminus B^*$. The choice of algorithm may depend on the structure of the matroid. Indeed, for graphic matroids, there are specialized algorithms that do not appear to extend to arbitrary matroids.

**Problem (Dual of a linear matroid).** Prove that if $[I, A]$ is a representation of a matroid $M$, then $[-A^T, I]$ is a representation of the dual matroid $M^*$.

**Exercise [Maximum-weight spanning tree, continued (see pp. 58, 60)].** With respect to the edge-weighted graph of the Maximum-weight spanning tree Exercise, use the Greedy Algorithm to find a *minimum*-weight base of the dual of the graphic matroid of the graph.

**Exercise [Scheduling, continued (see p. 59)].** With respect to the Scheduling Exercise, use the Greedy Algorithm to find a *minimum*-weight base of the associated dual matroid.

**Problem (Cocircuits and coloops).** Let $M$ be the graphic matroid of a graph $G$. Describe the circuits of $M^*$ in terms of $G$. In particular, describe the loops of $M^*$ in terms of $G$.

A *planar embedding* of a graph $G$ is a drawing of $G$ in the plane with no edges crossing. With respect to a planar embedding of $G$, we construct the planar dual $G^*$ by having a vertex corresponding to each region of the planar embedding of $G$ and having an edge consisting of each pair of regions that share a common edge. Note that $G^*$ has a vertex corresponding to the outer region of the planar embedding of $G$. Evidently $G^*$ is planar as well, and it is easily drawn on top of the planar embedding of $G$. As each edge of $G^*$ naturally crosses an edge of $G$ in the pair of planar embeddings, it is natural to label each edge of $G^*$ with the label of $G$ corresponding to the edge that it crosses.

**Example (Planar dual).** Consider the planar graph $G$:

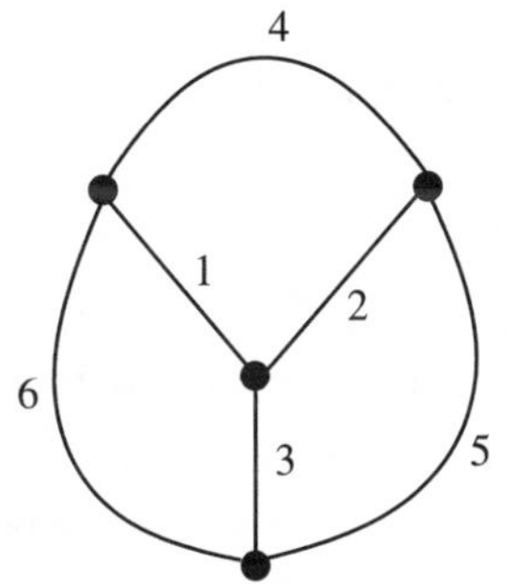

We construct the planar dual as the graph $G^*$, shown in the following figure with the hollow vertices and dashed edges:

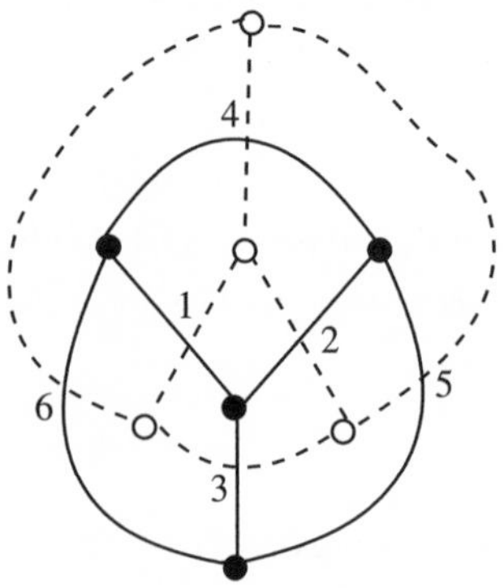


♠

**Problem (Dual graphic matroids and planar graphs).** Let $G$ be a planar graph. Take any planar embedding of $G$ and form the planar dual $G^*$. Prove that the graphic matroid of $G^*$ is the dual of the graphic matroid of $G$.

**Problem (Minors of matroids).** For a set $F \subset E(M)$, define $M \backslash F$ (read $M$ *delete* $F$) by $E(M \backslash F) := E(M) \setminus F$, and $\mathcal{I}(M \setminus F) := \{X \subset E(M) \setminus F : X \in \mathcal{I}(M)\}$. Clearly, $M \backslash F$ is a matroid. Now, define the matroid $M/F$ (read $M$ *contract* $F$) by $M/F := (M^* \backslash F)^*$.

a. Show that $r_{M/F}(X) = r_M(X \cup F) - r_M(F)$, $\forall\, X \subset E(M) \setminus F$.
b. Choose $J_F \subset F$ such that $J_F \in \mathcal{I}(M)$ and $|J_F| = \max\{|J| : J \subset F,\ J \in \mathcal{I}(M)\}$. Show that $\mathcal{I}(M/F) = \{X \subset E(M) \setminus F : X \cup J_F \in \mathcal{I}(M)\}$.
c. Describe how to obtain a representation of $M \backslash F$ and of $M/F$ from a representation of $M$.

## 1.7 The Matroid Polytope

The rank function leads to an appealing characterization of the independent sets of a matroid $M$ in terms of the extreme points of a polytope. Recall that

$$\mathcal{P}_{\mathcal{I}(M)} := \mathrm{conv}\{x(S) : S \in \mathcal{I}(M)\}.$$

**Theorem (Matroid polytope).** *For any matroid $M$,*

$$\mathcal{P}_{\mathcal{I}(M)} = \left\{ x \in \mathbf{R}_+^{E(M)} \ : \ \sum_{e \in T} x_e \le r_M(T), \quad \forall\, T \subset E(M) \right\}.$$

*Proof.* For every $S, T \subset E(M)$, we have

$$\sum_{e \in T} x_e(S) = |S \cap T| \,.$$

If $S \in \mathcal{I}(M)$, then $|S \cap T| \le r_M(T)$, as $S \cap T \subset T$ and $S \cap T \in \mathcal{I}(M)$. Therefore, by convexity, we have $\sum_{e \in T} x_e \le r_M(T)$ for all $x \in \mathcal{P}_{\mathcal{I}(M)}$, and we have

$$\mathcal{P}_{\mathcal{I}(M)} \subset \left\{ x \in \mathbf{R}_+^{E(M)} \ : \ \sum_{e \in T} x_e \le r_M(T), \quad \forall\, T \subset E(M) \right\}.$$

Hence, it suffices to show that *every* linear-objective function is maximized over

$$\left\{ x \in \mathbf{R}_+^{E(M)} \ : \ \sum_{e \in T} x_e \le r_M(T), \quad \forall\, T \subset E(M) \right\}$$

by a point of $\mathcal{P}_{\mathcal{I}(M)}$. Without loss of generality, let $E(M) = \{1, 2, \ldots, n\}$, and let $c(1) \ge c(2) \ge \cdots \ge c(n)$. Let $k$ be the greatest index among the nonnegative weights. Let $T_e := \{1, 2, \ldots, e\}$ for $1 \le e \le n$, and let $T_0 = \emptyset$. The Greedy Algorithm for finding a maximum-weight independent set $S$ can be viewed as determining its characteristic vector $x(S)$ as

$$x_e(S) := \begin{cases} r_M(T_e) - r_M(T_{e-1}), & \text{if } 1 \le e \le k \\ 0, & \text{if } k < e \le n \end{cases}.$$

The point $x(S)$ is a feasible solution of the linear program

$$(P) \qquad \begin{aligned} \max \ & \sum_{e \in E(M)} c(e) x_e \\ & \text{subject to:} \\ & \qquad \sum_{e \in T} x_e \le r_M(T), \quad \forall\, T \subset E(M); \\ & \qquad\quad x_e \ge 0, \quad \forall\, e \in E(M). \end{aligned}$$

We can check the feasibility of $x(S)$ by using only properties of $r_M$. Nonnegativity follows from R2. Satisfaction of the rank inequalities follows from R1–R3:

$$\begin{aligned}
\sum_{e \in T} x_e(S) &= \sum_{\substack{e \in T: \\ 1 \le e \le k}} \left( r_M(T_e) - r_M(T_{e-1}) \right) \\
&\le \sum_{\substack{e \in T: \\ 1 \le e \le k}} \left( r_M(T_e \cap T) - r_M(T_{e-1} \cap T) \right) \qquad \text{(by R3)} \\
&= r_M(T_k \cap T) - r_M(\emptyset) \\
&\le r_M(T) - r_M(\emptyset) \qquad \text{(by R2)} \\
&= r_M(T) \qquad \text{(by R1)}.
\end{aligned}$$

The dual of $P$ is the linear program

$$(D) \qquad \begin{aligned}
\min \sum_{T \subset E(M)} r_M(T)\, y_T \\
\text{subject to:} \\
\sum_{T : e \in T} y_T \ge c(e), \quad \forall\, e \in E(M) \\
y_T \ge 0, \quad \forall\, T \subset E(M).
\end{aligned}$$

As for $P$, we can construct a potential solution $y \in \mathbf{R}^{2^{E(M)}}$ of $D$, defined by

$$y_T := \begin{cases} c(e) - c(e+1), & \text{if } T = T_e \text{ with } 1 \le e < k \\ c(k), & \text{if } T = T_k \\ 0, & \text{otherwise} \end{cases} .$$

We need check only that $y$ is feasible to $D$ and that the objective value of $x(S)$ in $P$ and that of $y$ in $D$ are equal. Then, by the Weak Duality Theorem, $x(S)$ is optimal in $P$. Therefore, every linear function is maximized over

$$\left\{ x \in \mathbf{R}_+^{E(M)} : \sum_{e \in T} x_e \le r_M(T), \quad \forall\, T \subset E(M) \right\}$$

by a point of $\mathcal{P}_{\mathcal{I}(M)}$.

Clearly, $y$ is nonnegative. For $1 \le e < k$, we have

$$\begin{aligned}\sum_{T : e \in T} y_T &= \sum_{l=e}^{k} y_{T_l} \\ &= c(k) + \sum_{l=e}^{k-1} (c(l) - c(l+1)) \\ &= c(k) + \sum_{l=e}^{k-1} c(l) - \sum_{l=e+1}^{k} c(l) \\ &= c(e),\end{aligned}$$

which is certainly $\ge c(e)$. For $e = k$, we have

$$\sum_{T : k \in T} y_T = y_{T_k} = c(k),$$

which is certainly $\ge c(k)$. For $e > k$, we have

$$\sum_{T : e \in T} y_T = 0,$$

which is certainly $\ge c(e)$, because $c(e) < 0$ for $e > k$. Therefore, the solution $y$ is feasible to $D$. Finally, we have equality of the objective values because

$$\begin{aligned}\sum_{T \subset E(M)} y_T r_M(T) &= c(k) r_M(T_k) + \sum_{l=1}^{k-1} (c(l) - c(l+1)) r_M(T_l) \\ &= \sum_{l=1}^{k} c(l) r_M(T_l) - \sum_{l=2}^{k} c(l) r_M(T_{l-1}) \\ &= \sum_{l=1}^{k} c(l) \big( r_M(T_l) - r_M(T_{l-1}) \big) \\ &= \sum_{l=1}^{n} c(l) x_l.\end{aligned}$$

■

**Exercise (Dual solution).** With respect to the edge-weighted graph of the Maximum-weight spanning tree Exercise (see p. 58), calculate the "dual solution" of the previous proof, and use it to verify optimality of the maximum-weight forest.

**Example [Vertex packing on a star, continued (see pp. 51, 53, 57, 61)].** Let $c(1) = 2$ and $c(2) = c(3) = \cdots = c(n) = 1$. Following the definition of $x(S)$ in the previous proof, $x_1(S) = 1, x_2(S) = 0, x_3(S) = x_4(S) = \cdots = x_n(S) = 1$, which picks out the *dependent* set $S = \{1, 3, 4, \ldots, n\}$ having weight $n$, whereas the maximum-weight independent set is $\{2, 3, \cdots, n\}$, which has weight $n-1$. ♠

The proof of the characterization of $\mathcal{P}_{\mathcal{I}(M)}$ for matroids $M$ can be used to establish a related result.

**Theorem (Greedy optimality for polymatroids).** *Let $r$ be a function on $E := \{1, 2, \ldots, n\}$ satisfying R2, R3, and $r(\emptyset) = 0$. Suppose that $c(1) \geq c(2) \geq \cdots \geq c(n)$. Let $k$ be the greatest index among the nonnegative weights. Then the* greedy solution $x \in \mathbf{R}^E$ *defined by*

$$x_e := \begin{cases} r(T_e) - r(T_{e-1}), & \text{if } 1 \leq e \leq k \\ 0, & \text{if } k < e \leq n \end{cases}$$

*for all $e \in E$ solves the linear program*

$$\max \sum_{e \in E} c(e) x_e$$

subject to:

$$\sum_{e \in T} x_e \leq r(T), \quad \forall\, T \subset E;$$

$$x_e \geq 0, \quad \forall\, e \in E.$$

*Furthermore, if $k = n$ and we drop the inequalities $x_e \geq 0$, $\forall\, e \in E$, then we can omit the hypothesis that $r$ satisfies R2.*

For an independence system $M$, a set $T \subset E(M)$ is *inseparable* if the only $U \subset T$ for which $r_M(T) = r_M(U) + r_M(T \setminus U)$ are $U = T$ and $U = \emptyset$. Rank inequalities for sets that are not inseparable are redundant because

$$\sum_{j \in T} x_j \leq r_M(T)$$

is the sum of

$$\sum_{j \in U} x_j \leq r_M(U)$$

and

$$\sum_{j \in T \setminus U} x_j \leq r_M(T \setminus U),$$

when $r_M(T) = r_M(U) + r_M(T \setminus U)$.

For an independence system $M$, a set $T \subset E(M)$ is *closed* if $r_M(T + e) = r_M(T)$ for no $e \in E(M) \setminus T$. If $M$ is a matroid, then for every $T \subset E(M)$ there is a unique maximal superset $\mathrm{cl}_M(T)$ of $T$, called the *closure* (or *span*) of $T$, such that $r_M(T) = r_M(\mathrm{cl}_M(T))$.

Rank inequalities for sets that are not closed are redundant because

$$\sum_{j \in T} x_j \le r_M(T)$$

is the sum of

$$\sum_{j \in \mathrm{cl}_M(T)} x_j \le r_M(T)$$

and

$$-x_j \le 0, \quad \forall\, j \in \mathrm{cl}_M(T) \setminus T.$$

**Theorem (Facets of a matroid polytope).** *If $M$ is a matroid and $\{f\} \in \mathcal{I}(M)$, $\forall\, f \in E(M)$, then the rank inequalities for nonempty sets that are closed and inseparable, together with nonnegativity, provide a minimal description of $\mathcal{P}_{\mathcal{I}(M)}$.*

*Proof.* Clearly $\mathcal{P}_{\mathcal{I}(M)}$ is full dimensional because the $|E(M)| + 1$ points

$$x(\emptyset) \cup \{x(\{e\}) \ : \ e \in E(M)\}$$

are affinely independent. Therefore, each facet-describing valid inequality is unique up to multiplication by a positive scalar.

Each inequality $x_e \ge 0$ is facet describing because the $|E(M)|$ points

$$x(\emptyset) \cup \{x(\{f\}) \ : \ f \in E(M) - e\}$$

are affinely independent.

Next, suppose that nonempty $T$ is closed and inseparable and consider

$$\mathcal{F}(T) := \mathcal{P}_{\mathcal{I}(M)} \cap \left\{ x \in \mathbf{R}^{E(M)} \ : \ \sum_{e \in T} x_e = r_M(T) \right\}.$$

Clearly $\mathcal{F}(T)$ is a nontrivial face of $\mathcal{P}_{\mathcal{I}(M)}$. We demonstrate that, up to multiplication by a positive scalar, the only linear inequality that describes $\mathcal{F}(T)$ is $\sum_{e \in T} x_e \le r_M(T)$. By the Unique Description Theorem, this will demonstrate that $\mathcal{F}(T)$ is a facet of $\mathcal{P}_{\mathcal{I}(M)}$.

Let

$$\mathcal{T} := \left\{ S \in \mathcal{I}(M) \ : \ |S \cap T| = r_M(T) \right\},$$

and let

$$\mathcal{X}(\mathcal{T}) := \{x(S) \ : \ S \in \mathcal{T}\} \subset \mathcal{F}(T).$$

Observe that $S \in \mathcal{T}$ if and only if

$$x(S) \in \mathcal{P}_{\mathcal{I}(M)} \cap \left\{ x \in \mathbf{R}^{E(M)} : \sum_{e \in T} x_e = r_M(T) \right\}.$$

Let $\sum_{e\in E(M)} \alpha_e x_e \le \beta$ be an arbitrary inequality that describes $\mathcal{F}(T)$. Therefore, all points $x \in \mathcal{X}(\mathcal{T})$ satisfy $\sum_{e\in E(M)} \alpha_e x_e = \beta$.

Let $J$ be a maximal independent subset of $T$. Clearly $J \in \mathcal{T}$, so

$$(*) \qquad \sum_{e\in E(M)} \alpha_e x_e(J) = \sum_{e\in J} \alpha_e = \beta.$$

Consider $f \in E(M) \setminus T$. Because $T$ is closed, we have $J + f \in \mathcal{I}(M)$; hence, $J + f \in \mathcal{T}$ and

$$(**) \qquad \sum_{e\in E(M)} \alpha_e x_e(J+f) = \sum_{e\in J+f} \alpha_e = \beta.$$

Subtracting $(*)$ from $(**)$, we get $\alpha_f = 0$ for $f \in E(M) \setminus T$.

Next, we demonstrate that $\alpha_e = \alpha_f$ for all distinct $e, f$ in $T$. The following figure may help. Suppose otherwise. Let $T_1 = \{e \in T \ : \ \alpha_e$ is maximized over $T\}$. Let $T_2 = T \setminus T_1$. Let $J_2$ be a maximal independent subset of $T_2$. Extend $J_2$ to a maximal independent subset $J$ of $T$. Let $J_1 = J \setminus J_2$. Because $T$ is inseparable we have $|J_1| < r_M(T_1)$ [notice that $r_M(T) = |J|$, $r_M(T_2) = |J_2|$]. Therefore, there is some $e \in T_1 \setminus J_1$ such that $J_1 + e \in \mathcal{I}(M)$. It follows that there is some $e' \in J_2$ such that $J' := J + e - e'$ is a maximal independent-subset of $T$ (notice that $J + e$ contains a unique circuit, and that circuit is contained in $J_2 + e$; so choose $e' \in J_2$ to be any element of that circuit). Now, $J$ and $J'$ are both in $\mathcal{T}$, but $\sum_{e\in J'} \alpha_e > \sum_{e\in J} \alpha_e$. Hence, $\sum_{e\in T} \alpha_e x_e(J') > \sum_{e\in T} \alpha_e x_e(J)$, which is a contradiction.

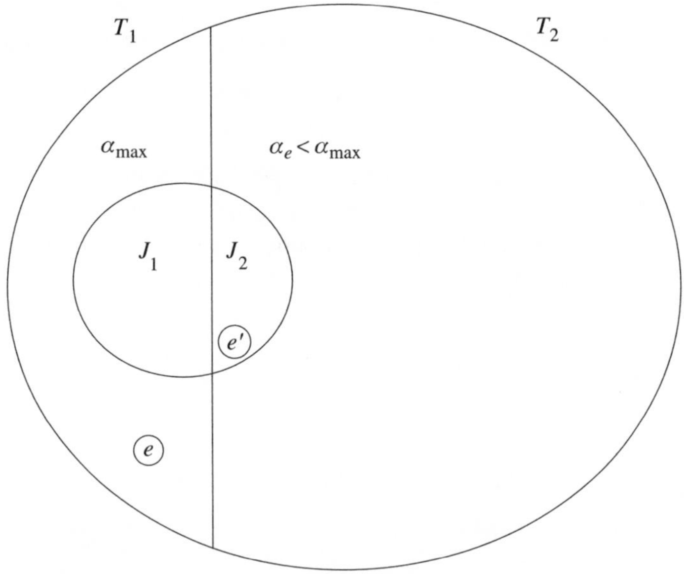

Therefore, every inequality describing $\mathcal{F}(T)$ has the form $\alpha \sum_{e \in T} x_e \leq \beta$. Plugging in $x(J)$ for some maximal independent subset of $T$ shows that $\beta = \alpha \cdot r_M(T)$. Finally, we find that the result follows by noting that (1) $\alpha = 0$ would imply $\mathcal{F}(T) = \mathcal{P}_{\mathcal{I}(M)}$, and (2) $\alpha < 0$ yields an inequality that is not valid. ■

**Problem (Base polytope).** Let $M$ be a matroid. Suppose that, for every pair of elements $g \neq h$, there is a circuit containing both. Let $\mathcal{P}_{\mathcal{B}(M)}$ be the convex hull of the characteristic vectors of *bases* of $M$.

a. Give one (nontrivial) linear equation satisfied by all points in $\mathcal{P}_{\mathcal{B}(M)}$.
b. Suppose that

$$\sum_{e \in E(M)} \alpha_e x_e = \beta$$

is an equation satisfied by all points in $\mathcal{P}_{\mathcal{B}(M)}$. Show that $\alpha_g = \alpha_h$ for every pair of elements $g \neq h$.
c. Show that $\dim(\mathcal{P}_{\mathcal{B}(M)}) = |E(M)| - 1$.
d. Give a complete description of $\mathcal{P}_{\mathcal{B}(M)}$ as the solution set of your equation from part a and additional linear inequalities.

**Problem (Base polytope with a coloop).** Let $M$ be a matroid. Suppose that $f$ is in *every* base of $M$. Suppose that, for every other pair of elements $g \neq h$ (both different from $f$), there is a circuit of $M$ containing $g$ and $h$.

a. Give two linearly independent equations satisfied by all points in $\mathcal{P}_{\mathcal{B}(M)}$.
b. Suppose that

$$\sum_{e \in E(M)} \alpha_e x_e = \beta$$

is an equation satisfied by all points in $\mathcal{P}_{\mathcal{B}(M)}$. Show that $\alpha_g = \alpha_h$ for every pair of elements $g \neq h$, both different from $f$.
c. Show that $\dim(\mathcal{P}_{\mathcal{B}(M)}) = |E(M)| - 2$.
d. Give a complete description of $\mathcal{P}_{\mathcal{B}(M)}$ as the solution set of your equations from part a and additional linear inequalities.

## 1.8 Further Study

The theory of matroids is a beautiful and deep area of combinatorial mathematics. The book by Oxley (1992) is a wonderful resource for learning about this subject.

There are many theoretical and practical studies of the application of greedy and local-search algorithms to combinatorial-optimization problems. One starting point is the book by Aarts and Lenstra (1997).

Chapter 13 of the book by Ahuja, Magnanti, and Orlin (1993) describes the details of efficient implementations of algorithms for the minimum-weight spanning tree problem.

# 2

## *Minimum-Weight Dipaths*

One of the simplest combinatorial-optimization problems is that of finding a minimum-weight dipath in an edge-weighted digraph (under some natural restrictions on the weight function). Not only are there rather simple algorithms for this problem, but algorithms for the minimum-weight dipath problem are fundamental building blocks for developing solution methods for more complicated problems.

Let $G$ be a strict digraph. A $v$–$w$ *diwalk* is a sequence of edges $e_i$, $1 \le i \le p$ (with $p \ge 0$), such that $t(e_1) = v$ (if $p > 0$), $h(e_p) = w$ (if $p > 0$), and $h(e_i) = t(e_{i+1})$, for $1 \le i < p$. Neither the edges nor the vertices need be distinct. The $v$–$w$ diwalk *imputes* the sequence of vertices $v = t(e_1), h(e_1) = t(e_2), h(e_2) = t(e_3), \ldots, h(e_{p-1}) = t(e_p), h(e_p) = w$. If no vertex in this imputed vertex sequence is repeated, then the $v$–$w$ diwalk is called a $v$–$w$ *dipath*. In such a case, every vertex of the imputed sequence other than $v$ and $w$ is called an *interior* vertex. Note that the empty sequence of edges is a $v$–$v$ diwalk for any vertex $v$; the associated imputed sequence of vertices is also empty, so the empty sequence of edges is a $v$–$v$ dipath. If $v = w$, and the only repetition in the imputed vertex sequence is the consonance of the first element with the last, then the diwalk is a *dicycle*. Therefore, the $v$–$w$ diwalk (with $v \ne w$) is a dipath if it contains no dicycle. A vertex $w$ is *accessible from* $v$ if there is a $v$–$w$ diwalk in $G$.

For a strict digraph $G$ and weight function $c$ on $E(G)$, we are interested in finding a minimum-weight $v$–$w$ dipath. If $w$ is not accessible from $v$, then there are no $v$–$w$ dipaths and the problem is infeasible. If $G$ contains no dicycle with negative weight, then any minimum-weight $v$–$w$ dipath is a minimum-weight $v$–$w$ diwalk. If $G$ contains a dicycle with negative weight, then there is some pair of vertices $v, w$ for which there are $v$–$w$ diwalks with weight less than any constant.

## 2.1 No Negative-Weight Cycles

Given a vertex $v$, the Bellman–Ford Algorithm calculates minimum-weight dipaths from $v$ to every other vertex. The algorithm will fail only if $G$ contains a diwalk from $v$ to some vertex $w$ that is contained in a negative-weight dicycle. In such a case, $w$ is accessible from $v$, but there is no minimum-weight $v$–$w$ diwalk.

The algorithm is based on the following definition. For $w \in V(G)$ and $0 \leq k \leq |V(G)| - 2$, let

$$f_k(w) := \text{weight of a minimum-weight } v\text{–}w \text{ diwalk with} \leq k \text{ interior vertices},$$

unless there is no $v$–$w$ diwalk with $\leq k$ interior vertices, in which case we define $f_k(w) := +\infty$. Note that $f_0(v) = 0$.

No $v$–$w$ dipath contains more than $|V(G)| - 2$ interior vertices; therefore, if $w$ is accessible from $v$ and $G$ contains no negative-weight dicycles, then

$$f_{|V(G)|-2}(w) := \text{weight of a minimum-weight } v\text{–}w \text{ dipath.}$$

The algorithm computes the numbers $f_k(w)$ for successively larger values of $k$, starting with $k = 0$.

**The Bellman-Ford Algorithm**

1. $f_0(v) := 0$, and
$$f_0(w) := \begin{cases} c((v, w)), & \text{if } (v, w) \in E(G) \\ +\infty, & \text{otherwise} \end{cases},$$
$\forall\, w \in V(G) - v$.
2. For $k = 1$ to $|V(G)| - 2$:
$$f_k(w) := \min\left(\{f_{k-1}(w)\} \cup \left\{f_{k-1}(t(e)) + c(e)\ : e \in \delta_G^-(w)\right\}\right),$$
$$\forall\, w \in V(G).$$

Because each edge is examined once for each $k$, the Bellman–Ford Algorithm requires $\mathcal{O}(|V(G)| \cdot |E(G)|)$ time.

**Exercise (Bellman–Ford Algorithm).** Use the Bellman–Ford Algorithm to find minimum-weight dipaths from vertex $a$ to all other vertices in the following digraph.

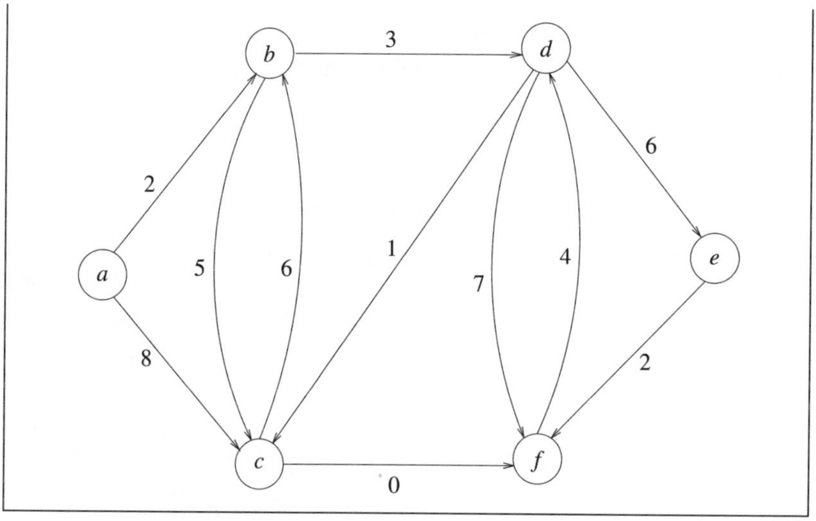


**Problem (Recovering the dipaths with the Bellman–Ford Algorithm).** Describe how minimum-weight $v$–$w$ dipaths can be recovered by keeping track of some extra information at each iteration of the Bellman–Ford Algorithm.

**Problem (Finding a negative-weight dicycle).** Suppose that for some $w \in V(G)$, we have $f_{|V(G)|-1}(w) < f_{|V(G)|-2}(w)$. Show that $G$ contains a $v$–$w$ diwalk that contains a negative-weight dicycle.

**Problem (Minimum-weight dipaths by linear programming).** Let $G$ be a digraph with $v \in V(G)$. Let $c$ be a weight function on $E(G)$ with the property that $G$ has no negative-weight dicycles. Consider the following linear program:

$$\max \sum_{e \in E(G)} c(e) x_e$$

subject to:

$$\sum_{e \in \delta_G^+(v)} x_e - \sum_{e \in \delta_G^-(v)} x_e = |V(G)| - 1 \,;$$

$$\sum_{e \in \delta_G^+(w)} x_e - \sum_{e \in \delta_G^-(w)} x_e = -1 \,, \quad \forall\, w \in V(G) - v \,;$$

$$x_e \geq 0 \,, \quad \forall\, e \in E(G) \,.$$

Demonstrate how to recover minimum-weight $v$–$w$ dipaths for all $w \in V(G) - v$ from an optimal solution of this linear program. Prove the correctness of your procedure.

## 2.2 All-Pairs Minimum-Weight Dipaths

If we want to calculate minimum-weight dipaths between all (ordered) pairs of vertices, we could just apply the Bellman–Ford algorithm $|V(G)|$ times, with each possible choice of a starting vertex $v$. This would require $\mathcal{O}(|V(G)|^4)$ time. The Floyd–Warshall Algorithm provides a way of calculating the same information in $\mathcal{O}(|V(G)|^3)$ time. Assume that the digraph $G$ contains no negative-weight dicycle. First we choose an arbitrary bijection $\pi : V(G) \mapsto \{1, 2, \ldots, |V(G)|\}$. For all ordered pairs of vertices $(v, w)$ and integers $k$ satisfying $0 \le k \le |V(G)|$, let

$$f_k(v, w) := \text{the weight of a minimum-weight } v\text{–}w \text{ dipath having all interior vertices } u \text{ satisfying } \pi(u) \le k.$$

**The Floyd–Warshall Algorithm**

1. $f_0(v, v) := 0, \quad \forall\, v \in V(G)$, and

$$f_0(v, w) := \begin{cases} c((v, w)), & \text{if } (v, w) \in E(G) \\ +\infty, & \text{otherwise} \end{cases},$$

$\forall\, w \in V(G) - v$.

2. For $k = 1$ to $|V(G)|$,

$$f_k(v, w) := \min\left\{ f_{k-1}(v, w),\ f_{k-1}(v, \pi^{-1}(k)) + f_{k-1}(\pi^{-1}(k), w) \right\},$$

$\forall\, v \ne w \in V(G)$.

**Problem (Recovering the dipaths with the Floyd-Warshall Algorithm).** Describe how minimum-weight $v$–$w$ dipaths can be recovered for all (ordered) pairs of vertices $v, w$ by keeping track of some extra information at each iteration of the Floyd–Warshall Algorithm.

## 2.3 Nonnegative Weights

There is another algorithm that is more efficient than the Bellman–Ford Algorithm, requiring just $\mathcal{O}(|V(G)|^2)$ time, but it requires that the weight function $c$

be nonnegative. In such a case, there is no possibility of negative-weight dicycles. Dijkstra's Algorithm maintains upper-bound labels $f(w)$ on the lengths of minimum-weight $v$–$w$ dipaths for all $w \in V(G)$. Throughout, the labels are partitioned into two classes: permanent and temporary. At each iteration of the algorithm, a temporary label that is least is made permanent, and the remaining temporary labels are updated. At any stage of the algorithm, the interpretation of the labels is as follows:

$$f(w) := \text{the weight of a minimum-weight } v\text{–}w \text{ dipath having all interior vertices permanently labeled.}$$

Initially the only permanent label is $f(v) := 0$. The other labels are all temporary.

**Dijkstra's Algorithm**

1. $P := \{v\}$. $f(v) := 0$. $T := V(G) - v$. For all $w \in T$,
$$f(w) := \begin{cases} c((v,w)), & \text{if } (v,w) \in E(G) \\ +\infty, & \text{otherwise} \end{cases}.$$
2. While $(T \neq \emptyset)$:
   i. choose $w^* \in T$ such that $f(w^*) = \min\{f(w) : w \in T\}$;
   ii. $T := T - w^*$, $P := P + w^*$;
   iii. for $e \in \delta_G^+(w^*)$ such that $h(e) \in T$,
   $f(h(e)) := \min\{f(h(e)), f(w^*) + c(e)\}$.

**Exercise (Dijkstra's Algorithm).** Use Dijkstra's Algorithm to find minimum-weight dipaths from vertex $a$ to all other vertices in the digraph from the Bellman–Ford Algorithm Exercise (see p. 76).

**Proposition (Correctness of labels for Dijkstra's algorithm).** *At the start of any iteration of Dijkstra's Algorithm, the following statements hold:*

a. *For all $w \in P$, $f(w) =$ the weight of a minimum-weight $v$–$w$ dipath;*
b. *For all $w \in T$, $f(w) =$ the weight of a minimum-weight $v$–$w$ dipath that has all interior vertices in $P$.*

*Proof.* The proof is by induction on $|P|$. The result is clear when $|P| = 1$. We assume that hypotheses $a$ and $b$ hold for the partition $P, T$. We want to verify that they continue to hold after we make the label of $w^*$ permanent and update the temporary labels. To make this precise, let $P' := P + w^*$, let $T' := T - w^*$, let $f'(w) := f(w)$ for $w \in P'$ and for $w \in T'$ such that $w \notin \delta_G^+(w^*)$, and let $f'(w) := \min\{f(w), f(w^*) + c((w^*, w))\}$ for $w \in T'$ such that $w \in \delta_G^+(w^*)$. We seek to verify subsequent hypotheses $a'$ and $b'$, assuming that $a$ and $b$ hold:

a$'$. For all $w \in P'$, $f'(w) =$ the weight of a minimum-weight $v$–$w$ dipath;
b$'$. For all $w \in T'$, $f'(w) =$ the weight of a minimum-weight $v$–$w$ dipath that has all interior vertices in $P'$.

First, we verify $a'$. Because, $f'(w) = f(w)$ for all $w \in P$ (by $a$), we need only to verify $a'$ for $w = w^*$. Consider any $v$–$w^*$ dipath $F$. Let $k$ be the first vertex in $T$ visited by $F$. Then we can think of $F$ as being composed of a $v$–$k$ dipath $F_1$ and a $k$–$w^*$ dipath $F_2$:

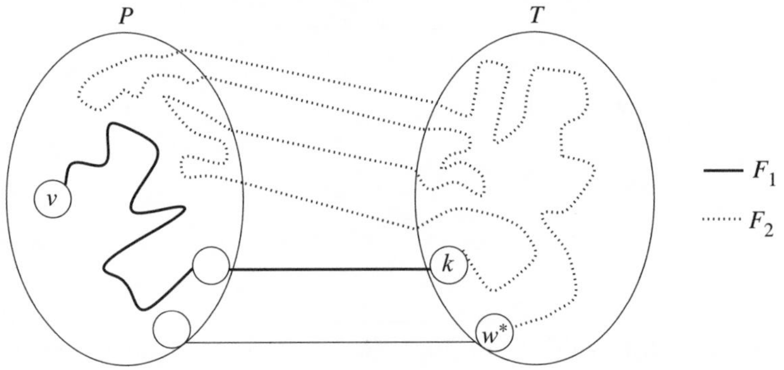


By the choice of $w^*$ in the algorithm, $f(k) \geq f(w^*)$. Furthermore, the weight of $F_2$ is nonnegative. By applying the inductive hypothesis $b$ to vertex $k$, we have that the weight of $F_1$ is at least $f(k)$. Therefore, the weight of $F$ is at least $f(w^*)$. Therefore, we have that no dipath using a vertex of $T$ as an interior vertex can have weight less than $f(w^*)$. Therefore, $a'$ holds.

Next, we verify $b'$ by using $a$. For $w \in T'$, consider a $v$–$w$ dipath $F'$ that has minimum weight among all $v$–$w$ dipaths that have their interior vertices in $P'$:

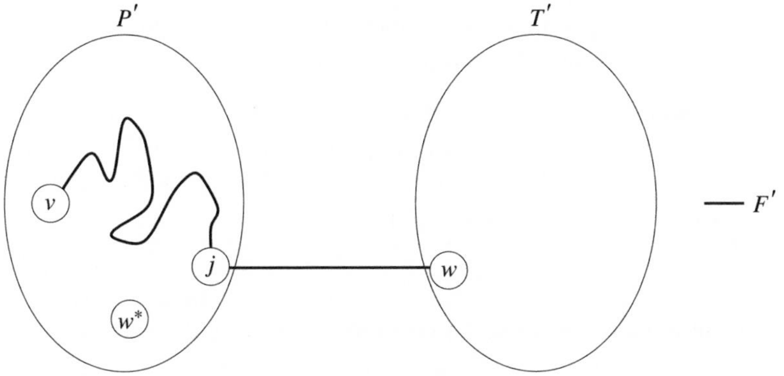


It cannot be that $w^*$ must be used by such an $F'$ before the last interior vertex $j$, because hypothesis $a$ implies that there is a minimum-weight $v$–$j$ dipath that does not use $w^*$. Therefore, we can choose $F'$ so that either the last interior vertex of $F'$ is $w^*$ or the dipath does not use $w^*$ at all. Then the definition of $f'(w)$ for $w \in T'$ ensures that hypothesis $b'$ holds. ■

**Corollary (Dijkstra's Theorem).** *At the conclusion of Dijkstra's Algorithm, $f(w)$ is the weight of a minimum-weight $v$–$w$ dipath for all $w \in V(G)$.*

**Problem (Recovering the dipaths with Dijkstra's Algorithm).** Describe how, in Dijkstra's Algorithm, we can recover a set of minimum-weight $v$–$w$ dipaths by having the algorithm maintain a "directed tree of dipaths rooted at $v$" at each iteration.

## 2.4 No Dicycles and Knapsack Programs

Rather than requiring the weight function $c$ to be nonnegative, another way to eliminate the possibility of negative-weight dicycles is to stipulate that the digraph have no dicycles whatsoever. Such an assumption allows a simple $\mathcal{O}(|E(G)|)$-time algorithm.

**Problem (Minimum-weight dipaths in graphs with no dicycles).** Suppose that $G$ has no dicycles. We can find a bijection $\pi : V(G) \mapsto \{1, 2, \ldots, |V(G)|\}$ so that for every $e \in E(G)$, $\pi(t(e)) < \pi(h(e))$.

$$f(w) := \text{the weight of a minimum-weight } v\text{–}w \text{ dipath having all vertices } u \text{ satisfying } \pi(u) \le \pi(w).$$

Starting with $f(\pi^{-1}(1)) = 0$, show how to compute $f(\pi^{-1}(k))$ for successively greater values of $k$, from $k = 1$ up through $k = |V(G)|$. Explain how this yields an $\mathcal{O}(|E(G)|)$-time algorithm for calculating minimum-weight $\pi^{-1}(1) - w$ dipaths for all $w \in V(G)$.

**Problem/Exercise (Knapsack program).** Consider the integer program

$$z := \max \sum_{j=1}^{n} c_j x_j$$

subject to:

$$\sum_{j=1}^{n} a_j x_j \le b;$$

$$x_j \ge 0, \quad j = 1, 2, \ldots n;$$

$$x_j \in \mathbf{Z}, \quad j = 1, 2, \ldots, n,$$

where $a_j$ and $b$ are positive integers.

a. Formulate the problem of calculating $z$ as a minimum-weight $v$–$w$ dipath problem (with $v$ fixed) on a digraph with no dicycles. *Hint:* The digraph should have $b + 1$ vertices. Try out your method on the following example:

$$\begin{array}{lrcrcrcrcrl} \text{max} & 11x_1 & + & 7x_2 & + & 5x_3 & + & x_4 & & & \\ \text{subject to:} & 6x_1 & + & 4x_2 & + & 3x_3 & + & x_4 & \le & 25; & \\ & x_1 & , & x_2 & , & x_3 & , & x_4 & \ge & 0 & \text{integer.} \end{array}$$

Hint: You can carry out the algorithm *without* drawing the digraph.

b. How can you change the general formulation if the $x_j$ are required to be integers between 0 and $u_j$? *Hint:* The digraph should have $1 + n(b + 1)$ vertices.

## 2.5 Further Study

A more detailed treatment of minimum-weight dipath problems is available in Chapters 4 and 5 of the book by Ahuja, Magnanti and Orlin (1993).

Most of the techniques of this chapter also fall in the domain of a subject called dynamic programming; Denardo (1982) is a standard reference in this field. Exceptionally, Dijkstra's Algorithm is not ordinarily considered to be a dynamic-programming algorithm. Ironically, the evolution of the set of permanently labeled vertices makes Dijkstra's Algorithm look much more *dynamic* than the static definitions of the dynamic-programming "value functions."

Cook and Seymour (2002) employ a very sophisticated and dynamic decomposition in their approach to combinatorial-optimization problems on sparse graphs.

# 3

## Matroid Intersection

Matroids become a particularly useful modeling tool in combinatorial optimization when we define more than one of them having a common ground set. Applications of this idea include the study of (1) bipartite matching, (2) the mechanics of frameworks, and (3) directed Hamiltonian tours. In particular, when the feasible solutions of a linear-objective combinatorial-optimization problem are sets that are independent in *two* matroids on a common ground set, striking optimization algorithms and polyhedral results apply.

### 3.1 Applications

For $p \geq 2$, let $M_i$ be matroids having the common ground set $E := E(M_i)$, $i = 1, 2, \ldots, p$, and let $c$ be a weight function on $E$. It is not generally the case that $\cap_{i=1}^{p} \mathcal{I}(M_i)$ is the set of independent sets of a matroid on $E$, even for $p = 2$. Therefore, a greedy algorithm is not appropriate for reliably calculating maximum-weight sets $S_k \in \cap_{i=1}^{p} \mathcal{I}(M_i)$ of all possible cardinalities $k$, even for $p = 2$. Indeed, a greedy algorithm can fail to deliver a maximum-*cardinality* set in $\cap_{i=1}^{p} \mathcal{I}(M_i)$, even for $p = 2$.

**Example (The intersection of two matroids need not be a matroid).** Let $M_1$ be the graphic matroid of the graph

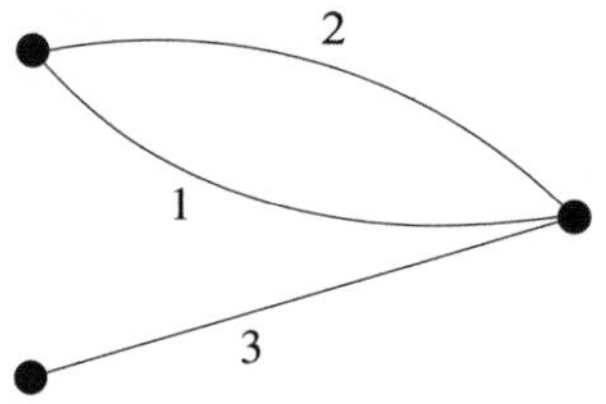

and let $M_2$ be the graphic matroid of the graph

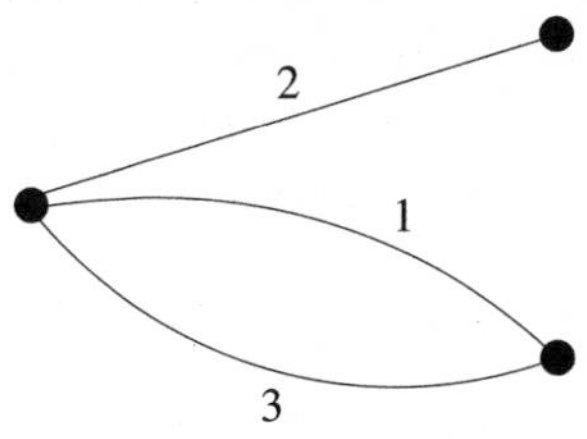


Therefore,

$$\mathcal{I}(M_1) \cap \mathcal{I}(M_2) = \{\emptyset, \{1\}, \{2\}, \{3\}, \{2, 3\}\}.$$

In fact, $\mathcal{I}(M_1) \cap \mathcal{I}(M_2)$ is the set of matchings of the bipartite graph

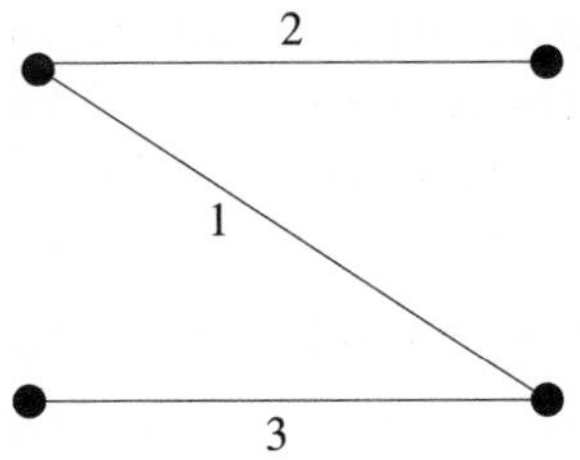


Now, if we try to build a maximum-cardinality element of $\mathcal{I}(M_1) \cap \mathcal{I}(M_2)$, one element at a time, in a myopic manner, we may fail. For example, if we take $S_0 := \emptyset$ and then $S_1 := \{1\}$, we cannot continue, even though there is a larger common independent set. ♠

In fact, the preceding example is an instance of a nice family of examples related to matchings in bipartite graphs.

**Example (Bipartite matching).** Let $G$ be a bipartite graph with vertex partition $V_1(G)$, $V_2(G)$ [that is, $V(G) = V_1(G) \cup V_2(G)$, $V_1(G) \cap V_2(G) = \emptyset$, $E(G[V_1]) = E(G[V_2]) = \emptyset$]. We define two matroids $M_1$ and $M_2$, having the common ground set $E(G)$, by

$$\mathcal{I}(M_i) := \{F \subset E(G) \, : |F \cap \delta_G(v)| \leq 1, \quad \forall\, v \in V_i(G)\},$$

for $i = 1, 2$. Clearly, $F \in \mathcal{I}(M_1) \cap \mathcal{I}(M_2)$ if and only if $F$ is a matching of $G$. ♠

**Example (Generic rigidity in the plane).** A *framework* $G$ consists of a finite set of points $V(G)$ in $\mathbf{R}^d$ and a set of straight lines $E(G)$ connecting some pairs of points. A *(infinitesimal) motion* of $G$ is an assignment of velocity vectors $m^v \in \mathbf{R}^d$, for all $v \in V(G)$, so that $v - w$ is perpendicular to $m^v - m^w$ whenever a line connects $v$ and $w$. That is,

$$\langle m^v - m^w, v - w\rangle = 0, \qquad \forall\, e = \{v, w\} \in E(G).$$

We can easily interpret the equations describing motions by rewriting them as

$$\frac{\langle m^v, v - w\rangle}{\|v - w\|} = \frac{\langle m^w, v - w\rangle}{\|v - w\|}, \qquad \forall\, e = \{v, w\} \in E(G).$$

In this form, we see that the equations dictate that the component of $m^v$ in the direction of the straight line connecting $v$ and $w$ should be the same as that of $m^w$.

Considering the scalar variables that are the components of the velocity vectors, we have a homogeneous system of $|E(G)|$ linear equations in $d|V(G)|$ variables, where $d$ is the dimension of the ambient Euclidean space:

$$\sum_{i=1}^{d} (v_i - w_i)m_i^v + (w_i - v_i)m_i^w = 0, \qquad \forall\, e = \{v, w\} \in E(G).$$

**Exercise (Motion).** Consider the following framework, in which the points are labeled with their Cartesian coordinates:

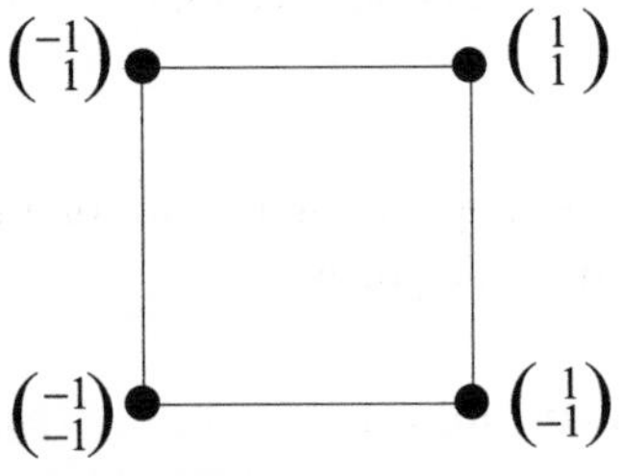


Write down the system of four linear equations in eight unknowns that describes the motions of this framework.

Every framework has some trivial motions – that is, those induced by the rigid motions of $\mathbf{R}^d$. We confine our attention to $d = 2$. The space of such rigid motions of the plane is three dimensional; for example, we can take as a basis horizontal translation, vertical translation, and clockwise rotation. Formally, we can realize a horizontal translation of points by using the velocity vector $m^v$ defined by

$$m^v := \begin{pmatrix} 1 \\ 0 \end{pmatrix}, \quad \forall\, v = \begin{pmatrix} v_1 \\ v_2 \end{pmatrix} \in V(G).$$

We can realize a vertical translation by using $m^v$ defined by

$$m^v := \begin{pmatrix} 0 \\ 1 \end{pmatrix}, \quad \forall\, v = \begin{pmatrix} v_1 \\ v_2 \end{pmatrix} \in V(G).$$

Finally, we can realize a clockwise rotation (about the origin) by using $m^v$ defined by

$$m^v := \begin{pmatrix} v_2 \\ -v_1 \end{pmatrix}, \quad \forall\, v = \begin{pmatrix} v_1 \\ v_2 \end{pmatrix} \in V(G).$$

**Exercise [Motion, continued (see p. 86)].** Find a nontrivial solution to the system of the Motion Exercise that does not correspond to a rigid motion of the plane.

Note that some motions are truly "infinitesimal." Consider the following framework, in which all vertices are held motionless except for one that is "moved" downward. This is not a true motion because any actual movement of just that one vertex in this framework is not possible. However, it is an infinitesimal motion. From an engineering point of view, it is quite practical to consider infinitesimal motions as motions because they do indicate an instability.

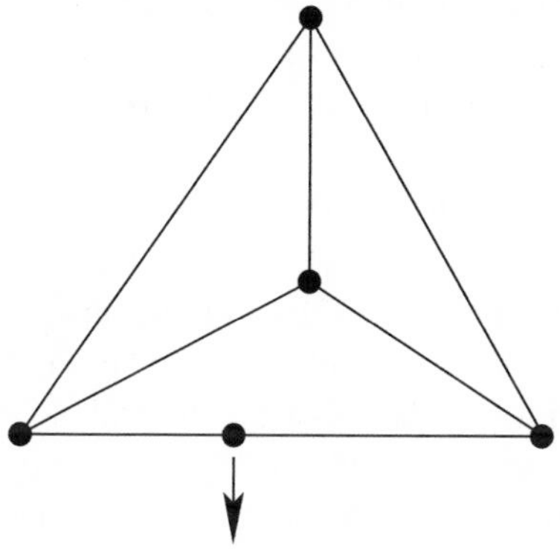

A framework is *infinitesimally rigid* (in the plane) if its only infinitesimal motions are rigid motions (of the plane). Equivalently, the framework is infinitesimally rigid if the rank of its associated linear system is $2|V(G)| - 3$. A framework is not infinitesimally rigid if $|E(G)| < 2|V(G)| - 3$. An infinitesimally rigid framework has unnecessary lines if $|E(G)| > 2|V(G)| - 3$. A framework is *minimally infinitesimally rigid* if it is infinitesimally rigid but ceases to be so if we delete any line.

A simple graph $G$ is *generically rigid (in the plane)* if it can be realized (in the plane) as an infinitesimally rigid framework with the lengths of its edges being algebraically independent over the rationals (i.e., the lengths should solve no polynomial equation having rational coefficients). For example, the preceding graph is generically rigid, and we see that by realizing it as the following framework:

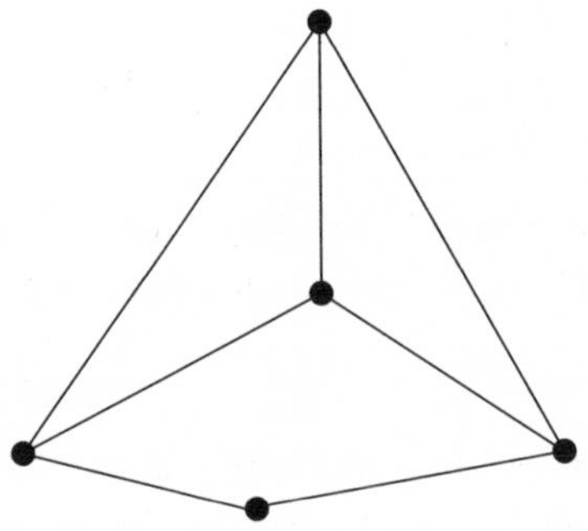

It turns out that there is a nice combinatorial characterization of which graphs $G$ that have $|E(G)| = 2|V(G)| - 3$ are generically rigid. For any $e \in E(G)$, let $G^e$ denote $G$ with the edge $e$ duplicated.

**Theorem (Planar generic rigidity).** *A simple graph $G$ is minimally generically rigid (in the plane) if $|E(G)| = 2|V(G)| - 3$ and $E(G^e)$ is the union of two (disjoint) spanning trees of $G^e$ for all $e \in E(G)$.*

For a proof and more, see Recski (1989) and Whiteley (1992).

For a graph $G$ having $|E(G)| = 2|V(G)| - 3$, we can test whether $E(G^e)$ is the union of two spanning trees of $G^e$ by considering the maximum cardinality of a set that is independent in a particular pair of matroids. Let $M_1$ be the graphic matroid of $G^e$, and let $M_2$ be the cographic matroid of $G^e$. Then $E(G^e)$ is the union of two spanning trees of $G^e$ if and only if there exists a set $S \in \mathcal{I}(M_1) \cap \mathcal{I}(M_2)$ with $|S| = |V(G^e)| - 1$. If there is such an $S$, then $S$ and $E(G^e) \setminus S$ are a pair of disjoint spanning trees of $G^e$. ♠

There are important examples arising from intersecting the independent sets of more than two matroids on a common ground set.

**Example (Directed Hamiltonian tours).** A *directed Hamiltonian tour* of digraph $G$ is a dicycle of $G$ that meets every vertex. We define three matroids $M_i$, $i = 1, 2, 3$, on the common ground set $E(G)$. First, we specify $M_1$ and $M_2$ by

$$\mathcal{I}(M_1) := \{F \subset E(G) \ : \ |F \cap \delta_G^+(v)| \le 1, \quad \forall\, v \in V(G)\}$$

and

$$\mathcal{I}(M_2) := \{F \subset E(G) \ : \ |F \cap \delta_G^-(v)| \le 1, \quad \forall\, v \in V(G)\}.$$

It is trivial to check that $M_1$ and $M_2$ are matroids (in fact, each is the direct sum of a collection of rank-1 and rank-0 uniform matroids). Next we choose an arbitrary vertex $w \in V(G)$, and, treating $G$ as an undirected graph, we let $M_3$ be the direct sum of the graphic matroid of $G[V(G) - w]$ and the uniform rank-2 matroid on $\delta_G(w)$. Then the edge sets of directed Hamiltonian tours of $G$ are precisely the sets in $\mathcal{I}(M_1) \cap \mathcal{I}(M_2) \cap \mathcal{I}(M_3)$ having cardinality $|V(G)|$. Indeed, $G$ has a directed Hamiltonian tour if and only if the maximum-cardinality elements of $\mathcal{I}(M_1) \cap \mathcal{I}(M_2) \cap \mathcal{I}(M_3)$ have $|V(G)|$ elements. ♠

## 3.2 An Efficient Cardinality Matroid-Intersection Algorithm and Consequences

In this section, it is shown how to efficiently find a maximum-cardinality element of $\mathcal{I}(M_1) \cap \mathcal{I}(M_2)$ for any pair of matroids $M_1, M_2$ with $E := E(M_1) = E(M_2)$. Before the algorithm is described, a few technical lemmata relating to matchings in bipartite graphs are established.

**Lemma (Unique matching implies crucial edge).** *Let $G$ be a bipartite graph with vertex bipartition $V_1(G)$, $V_2(G)$. Suppose that $G$ has a unique matching $X$ that meets all of $V_1(G)$. Then there exists an edge $e := \{v_1, v_2\} \in X$, with $v_1 \in V_1(G)$ and $v_2 \in V_2(G)$ such that*

$$\{v_1, v_2'\} \notin E(G), \ \forall\, v_2' \in V_2(G) - v_2.$$

*Proof.* The proof is by contradiction. If such $v_1$, $v_2$ do not exist, then there is a set $Y$ of edges extending between $V_1(G)$ and $V_2(G)$, with $|Y| = |X|$, $X \cap Y = \emptyset$, and with the property that $Y$ meets each element of $V_1(G)$ exactly once (see the following figure):

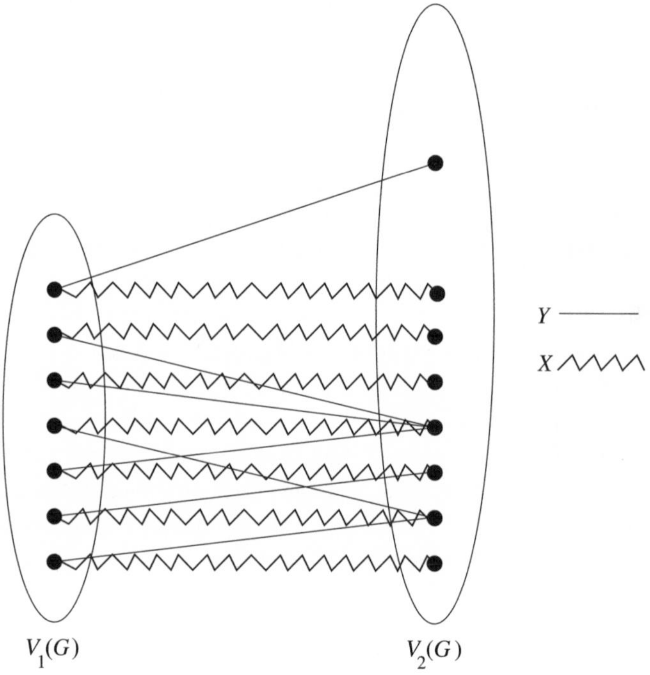


The set $X \cup Y$ must contain a nonempty path or cycle $C$ with an even number of edges, alternating between elements of $X$ and $Y$ *[just start walking along elements of $X \cup Y$, starting on the $V_2(G)$ side, first along an edge in $X$; after each edge in $X$, there is exactly one edge in $Y$ with which to continue the walk; after each edge in $Y$, there is at most one choice of edge in $X$ with which to continue the walk; eventually, we either (1) revisit a vertex in $V_2(G)$ closing a cycle, or (2) we reach a vertex in $V_2(G)$ that is not met by $X$, completing a path]*. Therefore, $X \Delta C$ is a matching that also meets $V_1(G)$, contradicting the uniqueness of $X$. ∎

Let $M$ be a matroid. With respect to any $S \in \mathcal{I}(M)$, we define the *bipartite exchange graph* $\mathcal{G}_M(S)$. The graph has $V(\mathcal{G}_M(S)) := E(M)$. All edges of $\mathcal{G}_M(S)$ extend between $S$ and $E(M) \setminus S$. Specifically, for $f \in S$ and $e \in E(M) \setminus S$,

$$\{f, e\} \in E(\mathcal{G}_M(S)) \text{ if } S - e + f \in \mathcal{I}(M);$$

that is, $\{f, e\} \in E(\mathcal{G}_M(S))$ if $S + e$ is independent, or, if not, if $f$ is in the unique circuit contained in $S + e$.

**Lemma (Exchange implies perfect matching).** *Let $M$ be a matroid with $S, T \in \mathcal{I}(M)$ and $|S| = |T|$. Then $\mathcal{G}_M(S)$ contains a perfect matching between $S \setminus T$ and $T \setminus S$.*

*Proof.* The proof is by contradiction. The subsequent figure helps in following the proof. Suppose that the hypothesis is true, but the conclusion is false. Then, Hall's Theorem (see p. 45) implies that there exists a set $W \subset T \setminus S$ such that $|N(W) \cap (S \setminus T)| < |W|$. Therefore, $(S \cap T) \cup W$ is a larger independent set than $(S \cap T) \cup [N(W) \cap (S \setminus T)]$. Hence, by I3, there exists an $e \in W$ such that

$$(S \cap T) \cup [N(W) \cap (S \setminus T)] + e \in \mathcal{I}(M).$$

Now, because $|S \setminus T| = |T \setminus S|$ and $|N(W) \cap (S \setminus T)| < |W|$, we must have $(S \setminus T) \setminus N(W) \neq \emptyset$. However, because there is no edge between $e$ and the nonempty set $(S \setminus T) \setminus N(W)$, it must be that $S + e \notin \mathcal{I}(M)$. Therefore, $S + e$ contains a unique circuit. However, because $(S \cap T) \cup [N(W) \cap (S \setminus T)] + e \in \mathcal{I}(M)$, that circuit must contain some $f \in (S \setminus T) \setminus N(W)$. However, then $\{f, e\}$ is an edge of $\mathcal{G}_M(S)$, in contradiction to the definition of $N(W)$.

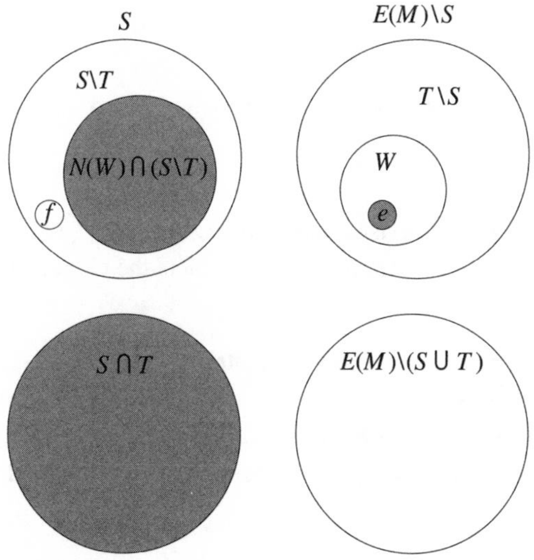


∎

A very useful partial converse holds.

**Lemma (Unique perfect matching implies exchange).** *Let $M$ be a matroid with $S \in \mathcal{I}(M)$. Suppose that $T \subset E(M)$, $|T| = |S|$, and $\mathcal{G}_M(S)$ contains a unique perfect matching $X$ between $S \setminus T$ and $T \setminus S$. Then $T \in \mathcal{I}(M)$.*

*Proof.* The proof is by induction on $|T \setminus S|$. The base case $|T \setminus S| = 0$ is trivial.

We assume that $|T \setminus S| \geq 1$. We apply the "Unique matching implies crucial edge" Lemma to the subgraph $G$ of $\mathcal{G}_M(S)$ induced by $S \Delta T$, with $V_1(G) := S \setminus T$ and $V_2(G) := T \setminus S$. Therefore, there exists an $f \in S \setminus T$ and an $e \in T \setminus S$ such that $\{f, e\} \in X$ and $\{f, e'\} \notin E(\mathcal{G}_M(S))$, for all $e' \in (T \setminus S) - e$. In particular, $S - f + e \in \mathcal{I}(M)$. Now, consider $T' := T - e + f$ and $X' := X - \{f, e\}$. Clearly $|T' \setminus S| < |T \setminus S|$, and $X'$ is the unique perfect matching in $\mathcal{G}_M(S)$ between $T' \setminus S$ and $S \setminus T'$. Therefore, by the inductive hypothesis, $T' \in \mathcal{I}(M)$; hence, by I2, $T - e = T' - f \in \mathcal{I}(M)$.

Therefore, by I3, there exists an $\tilde{e} \in (S - f + e) \setminus (T - e)$ such that $T - e + \tilde{e} \in \mathcal{I}(M)$. We may as well assume that $\tilde{e} \neq e$, because if $\tilde{e} = e$ we would conclude that $T \in \mathcal{I}(M)$, and we would be done. Hence, we may assume that there exists an $\tilde{e} \in (S - f) \setminus (T - e)$ such that $T - e + \tilde{e} \in \mathcal{I}(M)$. Therefore,

$$\begin{aligned} r_M((S \cup T') - f) &= r_M((S - f) \cup (T - e)) \qquad \left(\text{by the definition of } T'\right) \\ &\geq r_M(T - e + \tilde{e}) \\ &= |T - e + \tilde{e}| \qquad [\text{because } T - e + \tilde{e} \in \mathcal{I}(M)] \\ &= |S|. \end{aligned}$$

Therefore, by I3, there exists an $e' \in [(S \cup T') - f] \setminus (S - f) = (T - e) \setminus S$ such that $S - f + e' \in \mathcal{I}(M)$. This contradicts the choice of $f, e$. ■

Next, we return to the subject of matroid intersection. Let $M_1$ and $M_2$ be a pair of matroids with $E := E(M_1) = E(M_2)$. With respect to any $S \in \mathcal{I}(M_1) \cap \mathcal{I}(M_2)$, we define a *bipartite augmentation digraph* $\mathcal{G}_{M_1,M_2}(S)$. The graph has $V(\mathcal{G}_{M_1,M_2}(S)) := E$. All edges of $\mathcal{G}_{M_1,M_2}(S)$ extend between $S$ and

$E \setminus S$. The edges from $S$ to $E \setminus S$ are precisely the edges of $\mathcal{G}_{M_1}(S)$, oriented from $S$ to $E \setminus S$, but we omit the edges $\{f, e\}$ such that $S + e \in \mathcal{I}(M_1)$. Similarly, the edges from $E \setminus S$ to $S$ are precisely the edges of $\mathcal{G}M_2(S)$, oriented from $E \setminus S$ to $S$, but we omit the edges $\{f, e\}$ such that $S + e \in \mathcal{I}(M_2)$.

Certain vertices in $E \setminus S$ are termed sources and sinks. A *source* (respectively, sink) of $\mathcal{G}_{M_1,M_2}(S)$ is an $e \in E \setminus S$ such that $S + e \in \mathcal{I}(M_1)$ [respectively, $S + e \in \mathcal{I}(M_2)$]. An $e$–$e'$ dipath is a *source–sink dipath* if $e$ is a source and $e'$ is a sink. We include the degenerate case of an $e$–$e$ dipath having no edges, where $e$ is both a source and a sink.

**Example [Generic rigidity in the plane, continued (see p. 86)].** Consider the following graph $G$:

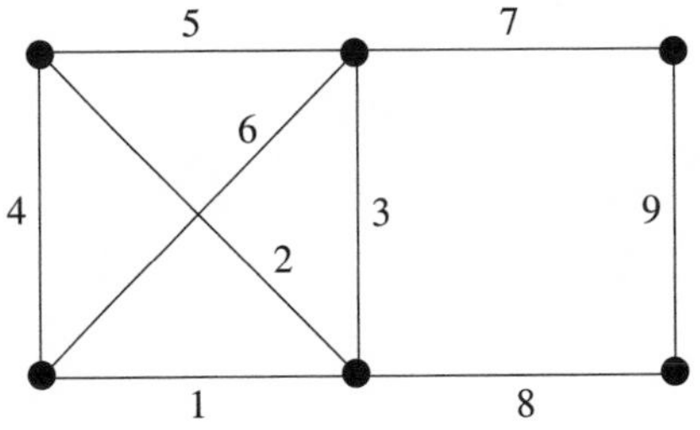


Notice that $|E(G)| = 9$ and $|V(G)| = 6$, so $G$ is a candidate for being a minimal generically rigid graph [i.e., $|E(G)| = 2|V(G)| - 3$]. Consider the graph $G^9$. Let edge 0 be the copy of edge 9. We seek to find a maximum-cardinality set that is independent in both $M_1$ and $M_2$. Consider the set $S := \{0, 1, 2, 3\} \in \mathcal{I}(M_1) \cap \mathcal{I}(M_2)$.

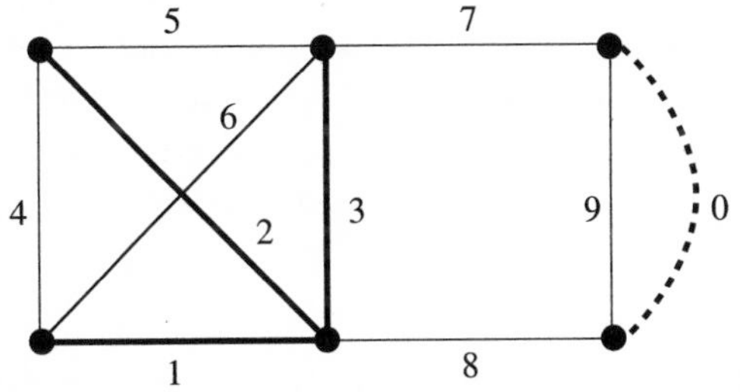

The *bipartite augmentation digraph* $\mathcal{G}(S)$ looks like

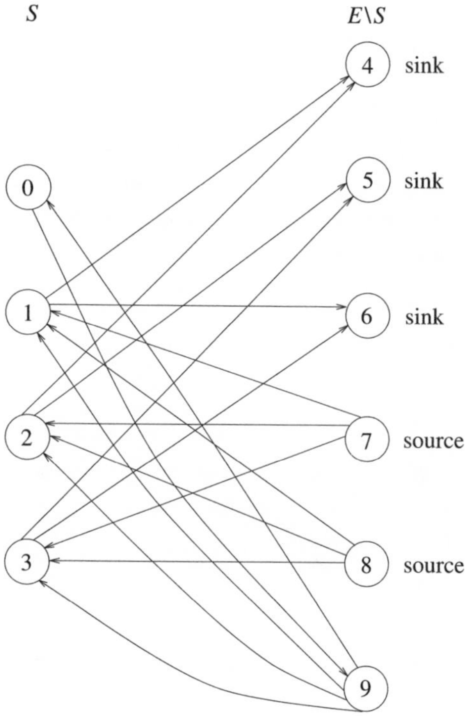


♠

With respect to a source–sink dipath $P$ in $\mathcal{G}_{M_1,M_2}(S)$, we have an imputed vertex sequence $e_0, f_1, e_1, f_2, e_2, \ldots, e_{n-1}, f_n, e_n$, where $e_0$ is a source, $e_n$ is a sink, all $e_i$ are in $E \setminus S$, all $f_i$ are in $S$, and all $(e_i, f_{i+1})$ and $(f_i, e_i)$ are edges of $\mathcal{G}_{M_1,M_2}(S)$. The source–sink dipath $P$ is *augmenting* if $S' := S \setminus \{f_1, f_2, \ldots, f_n\} \cup \{e_0, e_1, \ldots, e_n\}$ is in $\mathcal{I}(M_1) \cap \mathcal{I}(M_2)$.

We are able to obtain an augmenting sequence from a shortest source–sink dipath in $\mathcal{G}_{M_1,M_2}(S)$. It is easy to find a shortest source–sink dipath by use of a "breadth-first search," starting from the sources. A shortest source–sink dipath has no "shortcuts" [i.e., there is no edge $(v, w)$, where $w$ follows $v$ in the imputed vertex sequence, but not immediately] as well as no sources and sinks as interior vertices.

**Lemma (Shortest implies augmenting).** *Let $M_1$ and $M_2$ be matroids with $E := E(M_1) = E(M_2)$ and $S \in \mathcal{I}(M_1) \cap \mathcal{I}(M_2)$. If $P$ is a shortest source–sink dipath in $\mathcal{G}_{M_1,M_2}(S)$, then its imputed vertex sequence is augmenting.*

*Proof.* Consider the graph $\mathcal{G}_{M_1}(S)$. The edges $X := \{\{f_1, e_1\}, \{f_2, e_2\}, \ldots, \{f_n, e_n\}\}$ form a perfect matching between $\{f_1, f_2, \ldots, f_n\}$ and $\{e_1, e_2, \ldots, e_n\}$ in $\mathcal{G}_{M_1}(S)$. In fact, $X$ is the *unique* perfect matching between $\{f_1, f_2, \ldots, f_n\}$ and $\{e_1, e_2, \ldots, e_n\}$ in $\mathcal{G}_{M_1}(S)$ *[because, if there were another one, then, in that one, some $f_i$ $(i = 1, 2, \ldots, n)$ would be matched to an $e_j$ $(j = 1, 2, \ldots, n)$ with $j > i$; such an edge would be a shorter source–sink dipath than $P$]*. Then, by the "Unique perfect matching implies exchange" Lemma,

$$\widetilde{S} := (S \setminus \{f_1, f_2, \ldots, f_n\}) \cup \{e_1, e_2, \ldots, e_n\} \in \mathcal{I}(M_1).$$

However, we are not quite done; we must demonstrate that $S' = \widetilde{S} + e_0 \in \mathcal{I}(M_1)$. We have

$$r_{M_1}(S \cup \{e_0, e_1, \ldots, e_n\}) \geq r_{M_1}(S + e_0) = |S| + 1 \qquad \text{(because } e_0 \text{ is a source)},$$

and

$$r_{M_1}(S \cup \{e_1, e_2, \ldots, e_n\}) = |S| \quad \text{(because } \{e_1, e_2, \ldots, e_n\} \text{ contains no source)}.$$

Therefore, $S' = \widetilde{S} + e_0 \in \mathcal{I}(M_1)$.

By symmetry, we have $S' \in \mathcal{I}(M_2)$ as well. ∎

**Exercise (Shortcut).** Let $M_1$ and $M_2$ be the graphic matroids of the graphs $G_1$ and $G_2$, respectively. Show that, for $S := \{2, 4\}$, there is a source–sink dipath that does not yield an augmenting sequence.

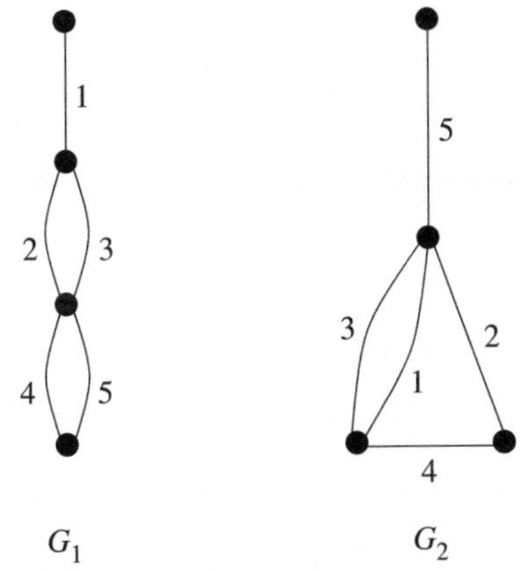

The "Shortest implies augmenting" Lemma suggests the following simple algorithm to find a maximum-cardinality set that is independent in $M_1$ and $M_2$.

**Cardinality Matroid-Intersection Algorithm**

1. Start with any $S \in \mathcal{I}(M_1) \cap \mathcal{I}(M_2)$. For example, $S := \emptyset$.
2. While $\mathcal{G}(S)$ has a source-sink dipath:
   i. let $e_0, e_1, f_1, e_2, f_2, \ldots, f_n, e_n$ be an augmenting sequence;
   ii. let $S := S \cup \{e_j : 0 \le j \le n\} \setminus \{f_j : 1 \le j \le n\}$.

**Example [Generic rigidity in the plane, continued (see pp. 86, 93)].** The bipartite augmentation digraph yields the augmenting sequence 8, 3, 6, so we are led to the set $\{0, 1, 2, 6, 8\} = \{0, 1, 2, 3\} \cup \{6, 8\} \setminus \{3\}$, which is in $\mathcal{I}(M_1) \cap \mathcal{I}(M_2)$.

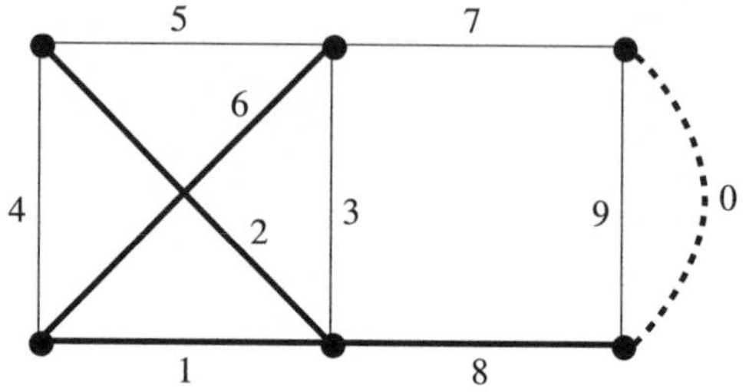


♠

**Theorem (Correctness of the Cardinality Matroid-Intersection Algorithm).** *On termination of the Cardinality Matroid-Intersection Algorithm, $S$ is a maximum-cardinality set in $\mathcal{I}(M_1) \cap \mathcal{I}(M_2)$.*

*Proof.* Suppose that $E = E_1 \cup E_2$. Then, for any $S \in \mathcal{I}(M_1) \cap \mathcal{I}(M_2)$,

$$|S| \le |S \cap E_1| + |S \cap E_2| \le r_{M_1}(E_1) + r_{M_2}(E_2).$$

Therefore, it suffices to find $E_1$ and $E_2$ that cover $E$ such that $|S| = r_{M_1}(E_1) + r_{M_2}(E_2)$.

Let

$$A_S := \{w \in E : \text{there is a } v\text{–}w \text{ dipath for some source } v \text{ of } \mathcal{G}(S)\}.$$

Let $E_1 := \mathrm{cl}_{M_1}((E \setminus A_S) \cap S)$ and $E_2 := \mathrm{cl}_{M_2}(A_S \cap S)$. Now,

$$\begin{aligned} r_{M_1}(E_1) + r_{M_2}(E_2) &= r_{M_1}((E \setminus A_S)) \cap S) + r_{M_2}(A_S \cap S) \\ &= |(E \setminus A_S) \cap S| + |A_S \cap S| = |S|. \end{aligned}$$

It remains to be shown that $E_1 \cup E_2 = E$. First, we make some simple observations. By the definition of $A_S$, (1) all of the sources are in $(E \setminus S) \cap A_S$, and all of the sinks are in $(E \setminus S) \cap (E \setminus A_S)$, and (2) there are no edges from $A_S \cap S$ to $(E \setminus S) \cap (E \setminus A_S)$ and no edges from $(E \setminus S) \cap A_S$ to $(E \setminus A_S) \cap S$.

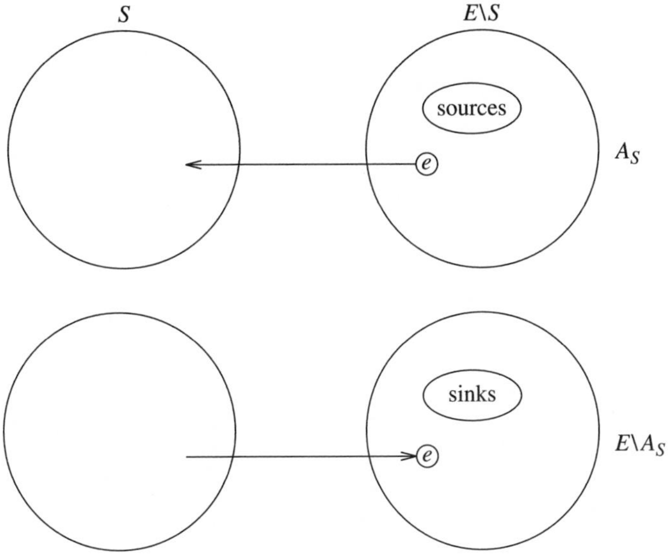


Clearly $e \in S$ implies that $e \in E_1 \cup E_2$. Therefore, suppose that $e \in E \setminus S$. If $e \in A_S$, then $e \in E_2$; otherwise $e$ would be a sink. If $e \in E \setminus A_S$, then $e \in E_1$; otherwise $e$ would be a source. ■

**Example [Generic rigidity in the plane, continued (see pp. 86, 93, 96)].** Now, consider $G^4$. This time, let edge 0 be the copy of edge 4. We seek to find a maximum-cardinality set that is independent in both $M_1$ and $M_2$. Consider the set $S := \{0, 3, 5, 7\}$.

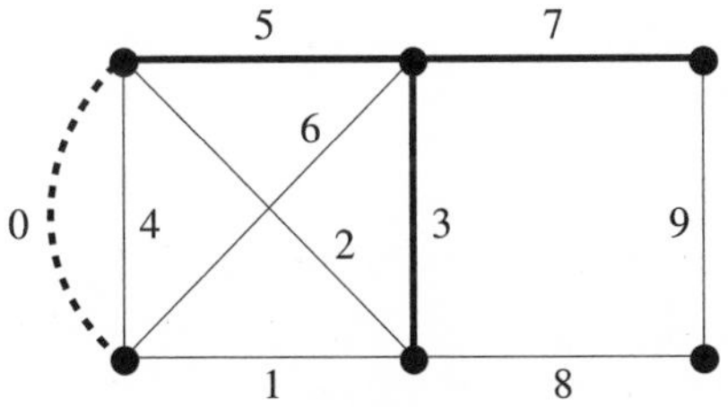

The bipartite augmentation digraph $\mathcal{G}(S)$ looks like

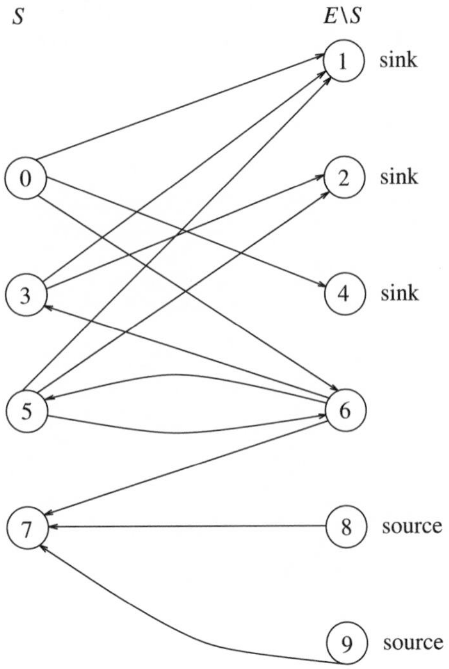


For this example, we have $A_S = \{7, 8, 9\}$. Hence,

$$\begin{aligned} E_1 &:= \mathrm{cl}_{M_1}((E \setminus A_S) \cap S) \\ &= \mathrm{cl}_{M_1}(\{0, 3, 5\}) \\ &= \{0, 1, 2, 3, 4, 5, 6\}, \end{aligned}$$

and

$$\begin{aligned} E_2 &:= \mathrm{cl}_{M_2}(A_S \cap S) \\ &= \mathrm{cl}_{M_2}(\{7\}) \\ &= \{7, 8, 9\}, \end{aligned}$$

Therefore, we have $E = E_1 \cup E_2$. Because $r_{M_1}(E_1) = 3$ and $r_{M_2}(E_2) = 1$, we have $|S| = 4 = r_{M_1}(E_1) + r_{M_2}(E_2)$. Therefore, $E(G^4)$ is not the disjoint union of two spanning trees. Hence, $G$ is not generically rigid. ♠

**Exercise (Generic rigidity in the plane).** Determine whether the following graph is generically rigid in the plane.

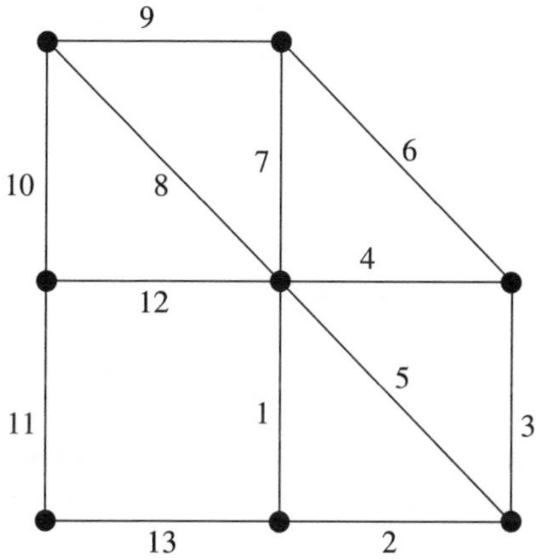


**Exercise [Scheduling, continued (see pp. 59, 65)].** Recalling the Scheduling Exercise, let $E := \{1, 2, \ldots, 10\}$, and let $M_1$ be the matroid having $E(M_1) := E$, and

$$\mathcal{I}(M_1) := \{X \subset E \ : \ X \text{ can be completed on time}\}.$$

Let $M_2$ be the matroid for which $E(M_2) := E$, and

$$\mathcal{I}(M_2) := \{X \subset E \ : \ |X \cap \{2i-1, 2i\}| \le 1, \ \text{for } i = 1, 2, 3, 4, 5\}.$$

Verify that the set $S := \{2, 6, 8, 9\}$ is in $\mathcal{I}(M_1) \cap \mathcal{I}(M_2)$. Construct the bipartite augmentation digraph $\mathcal{G}(S)$, and identify either a shortcut-free source–sink dipath or sets $E_1$, $E_2$ such that $E_1 \cup E_2 = E$ and $r_{M_1}(E_1) + r_{M_2}(E_2) = |S|$.

A consequence of the proof of the validity of the Cardinality Matroid-Intersection Algorithm is the following duality theorem.

**Matroid-Intersection Duality Theorem.**

$$\max\{|S| \ : \ S \in \mathcal{I}(M_1) \cap \mathcal{I}(M_2)\} = \min\left\{r_{M_1}(E_1) + r_{M_2}(E_2) \ : \ E_1 \cup E_2 = E\right\}.$$

**Example [Bipartite matching, continued (see p. 85)].** For $i = 1, 2$, $r_{M_i}(E_i)$ is precisely the number of elements of $V_i(G)$ that are met by $E_i$. Therefore, a consequence of the Matroid-Intersection Duality Theorem is König's famous characterization of maximum-cardinality matchings in bipartite graphs (see p. 44): The number of edges in a maximum-cardinality matching in a bipartite graph is equal to the minimum number of vertices needed to cover all of the edges of the graph. ♠

**Example (Separations).** A $k$-separation of matroid $M$ is a partition $(S_1, S_2)$ of $E(M)$ so that $|S_1| \geq k$, $|S_2| \geq k$, and

$$r_M(S_1) + r_M(S_2) \leq r_M(E(M)) + k - 1.$$

If $A$ is a representation of $M$ and $M$ has a $k$-separation $(S_1, S_2)$, then there is a nonsingular matrix $B$ and a permutation matrix $\Pi$ such that

$$BA\Pi = \begin{array}{c} \begin{matrix} S_1 & S_2 \end{matrix} \\ \begin{pmatrix} A_1 & 0 \\ C_1 & C_2 \\ 0 & A_1 \end{pmatrix} \end{array},$$

where $(C_1 \quad C_2)$ has $k - 1$ rows.

Now suppose that $X_1$ and $X_2$ are disjoint subsets of $E(M)$, each having cardinality $k$. We may consider $k$-separations $(S_1, S_2)$ such that $X_1 \subset S_1$ and $X_2 \subset S_2$. Letting $E_1 := S_1 \setminus X_1$ and $E_2 := S_2 \setminus X_2$ and using the formula for the rank in minors, we can reexpress the separation inequality as

$$r_{M/X_1\backslash X_2}(E_1) + r_{M/X_2\backslash X_1}(E_2) \leq r_M(E(M)) - r_M(X_1) - r_M(X_2) + k - 1.$$

Therefore, $M$ has a $k$-separation $(S_1, S_2)$ with $X_1 \subset S_1$ and $S_2 \subset E_2$ if and only if all common independent sets of $M/X_1\backslash X_2$ and $M/X_2\backslash X_1$ have cardinality less than $r_M(E(M)) - r_M(X_1) - r_M(X_2) + k$. By allowing $X_1$ and $X_2$ to vary, we can determine whether $M$ has a $k$-separation by solving at most $\mathcal{O}(|E(M)|^{2k})$ (cardinality) matroid-intersection problems. ♠

**Problem (Matroid partitioning).** Let $M_i$ be matroids on the common ground set $E$ for $i = 1, 2, \ldots, p$. Define an independence system $M$ such that $E(M) := E$ and

$$\mathcal{I}(M) := \left\{ S \subset E \; : \; S = \bigcup_{i=1}^{p} S_i, \; S_i \in \mathcal{I}(M_i), i = 1, 2, \ldots, p \right\}.$$

Prove that $M$ is a matroid by showing that

a. $r_M$ is defined by

$$r_M(X) = \min\left\{\sum_{i=1}^{p} r_{M_i}(T) + |X \setminus T| \;:\; T \subset X\right\}, \qquad \forall\, X \subset E,$$

and

b. $r_M$ satisfies R1–R3.

*Hint:* For part a, construct matroids $\overline{M}_X$ and $\widehat{M}_X$ on the common ground set

$$\widetilde{E}_X := X \times \{1, 2, \ldots, p\},$$

by letting

i. $\mathcal{I}(\overline{M}_X) :=$ {subsets of $\widetilde{E}_X$ such that no two elements have the same first component}, and
ii. $\mathcal{I}(\widehat{M}_X) :=$ {subsets of $\widetilde{E}_X$ such that the set of first components that have the same second component is in $\mathcal{I}(M_i)$, $i = 1, 2, \ldots, p$}.

Think of $\overline{M}_X$ as a matroid that permits partial $p$-colorings of $X$. Matroid $\widehat{M}_X$ forces elements of color $i$ to be independent in $M_i$, $i = 1, 2, \ldots, p$. Now, consider a maximum-cardinality element of $\mathcal{I}(\overline{M}_X) \cap \mathcal{I}(\widehat{M}_X)$.

## 3.3 An Efficient Maximum-Weight Matroid-Intersection Algorithm

With respect to matroids $M_1$, $M_2$ on the common ground set $E$ and weight function $c$, we consider the problem of finding maximum-weight sets $S_k$ of cardinality $k$ in $\mathcal{I}(M_1) \cap \mathcal{I}(M_2)$, for all $k$ for which such sets exist. Our algorithm is motivated by the algorithm for the cardinality case. The algorithm works by computing the desired $S_k$ for successively larger values of $k$, starting with $k = 0$ and $S_0 = \emptyset$.

As for the cardinality case, we work with the bipartite augmentation digraph. In the algorithm, if there is a sink that is accessible from a source in the bipartite augmentation digraph $\mathcal{G}(S_k)$, we augment by using the imputed vertex sequence of a certain dipath in $\mathcal{G}(S_k)$.

Let $e_1, f_1, e_2, f_2, \ldots, f_{n-1}, e_n$ be an augmenting sequence. Its *incremental weight* is

$$\sum_{j=1}^{n} c(e_j) - \sum_{j=1}^{n-1} c(f_j),$$

and its *length* is $n$.

**(Weighted) Matroid-Intersection Algorithm**

1. Start with $k$ and $S_k$ such that $S_k$ is a maximum-weight set in $\mathcal{I}(M_1) \cap \mathcal{I}(M_2)$. For example, let $k := 0$ and $S_0 := \emptyset$.
2. While $\mathcal{G}(S_k)$ has a source-sink dipath:
   i. let $e_1, f_1, e_2, f_2, \ldots, f_{n-1}, e_n$ be a shortest (length) augmenting sequence among those having maximum weight;
   ii. let $S_{k+1} := S_k \cup \{e_j : 1 \leq j \leq n\} \setminus \{f_j : 1 \leq j \leq n-1\}$;
   iii. let $k \leftarrow k+1$.

We note that it is not hard to find a shortest (length) augmenting sequence among those having maximum weight. This amounts to finding a minimum-weight source–sink dipath and (possibly repeatedly) checking whether there is any shortcut leading to a dipath with the same weight.

At termination of the algorithm, we claim that there is no $S_{k+1} \in \mathcal{I}(M_1) \cap \mathcal{I}(M_2)$ having cardinality $k+1$. This is easily verified in exactly the same manner as for the Cardinality Matroid-Intersection Algorithm.

Therefore, the only thing to verify is that, after each iteration, $S_{k+1}$ is a maximum-weight set of cardinality $k+1$ in $\mathcal{I}(M_1) \cap \mathcal{I}(M_2)$. Verification of this (nontrivial) fact is left to the industrious reader.

**Exercise [(Weighted) Matroid-Intersection Algorithm].** Consider the pair of matroids from the Shortcut Exercise (see p. 95). We define a weight function $c$ by

| $e$ | $c(e)$ |
|---|---|
| 1 | 5 |
| 2 | 4 |
| 3 | 7 |
| 4 | 8 |
| 5 | 9 |

We claim that $S_2 := \{3, 4\}$ is a maximum-weight set in $\mathcal{I}(M_1) \cap \mathcal{I}(M_2)$ having cardinality 2. Starting with $S_2$, use the (Weighted) Matroid-Intersection Algorithm to find a maximum-weight set $S_3 \in \mathcal{I}(M_1) \cap \mathcal{I}(M_2)$ having cardinality 3.

## 3.4 The Matroid-Intersection Polytope

Next, we establish an appealing characterization of the elements of $\mathcal{I}(M_1) \cap \mathcal{I}(M_2)$ in terms of the extreme points of a polytope. Recall that

$$\mathcal{P}_{\mathcal{I}(M_1)\cap\mathcal{I}(M_2)} := \mathrm{conv}\{x(S) \ : \ S \in \mathcal{I}(M_1) \cap \mathcal{I}(M_2)\}.$$

**Theorem (Matroid-Intersection Polytope).** *For any pair of matroids $M_1$, $M_2$, with common ground set $E$,*

$$\mathcal{P}_{\mathcal{I}(M_1)\cap\mathcal{I}(M_2)} = \mathcal{P}_{\mathcal{I}(M_1)} \cap \mathcal{P}_{\mathcal{I}(M_2)}.$$

*Proof.* The extreme points of $\mathcal{P}_{\mathcal{I}(M_1)\cap\mathcal{I}(M_2)}$ are the points $x(S) \in \mathbf{R}^E$ such that $S \in \mathcal{I}(M_1)$ and $S \in \mathcal{I}(M_2)$. Therefore, the extreme points of $\mathcal{P}_{\mathcal{I}(M_1)\cap\mathcal{I}(M_2)}$ lie in $\mathcal{P}_{\mathcal{I}(M_1)}$ and $\mathcal{P}_{\mathcal{I}(M_2)}$. Hence, $\mathcal{P}_{\mathcal{I}(M_1)\cap\mathcal{I}(M_2)} \subset \mathcal{P}_{\mathcal{I}(M_1)} \cap \mathcal{P}_{\mathcal{I}(M_2)}$.

We demonstrate the reverse inclusion by induction on $|E|$. The theorem is easily checked for $|E| = 1$. Therefore, suppose that $|E| > 1$. Let $z$ be an arbitrary extreme point of $\mathcal{P}_{\mathcal{I}(M_1)} \cap \mathcal{P}_{\mathcal{I}(M_2)}$. It is sufficient to prove that $z$ is 0/1 valued, as that would imply that $z \in \mathcal{P}_{\mathcal{I}(M_1)\cap\mathcal{I}(M_2)}$. First, we demonstrate that $z$ has at least one component that is 0 or 1.

Toward that goal, we may assume that $z_e > 0$ for all $e \in E$ (otherwise we would be done). For $i = 1, 2$, let

$$\mathcal{T}_i := \left\{ T \subset E \ : \ \sum_{e \in T} z_e = r_{M_i}(T) \right\}.$$

These sets pick out "tight rank inequalities" for each of the two matroid polytopes, with respect to the point $z$.

R3 implies that for $T, T' \in \mathcal{T}_i$, we have

$$\begin{aligned} r_{M_i}(T) + r_{M_i}(T') &\geq r_{M_i}(T \cap T') + r_{M_i}(T \cup T') \\ &\geq \sum_{e \in T \cap T'} z_e + \sum_{e \in T \cup T'} z_e \\ &= \sum_{e \in T} z_e + \sum_{e \in T'} z_e, \end{aligned}$$

so we have equality throughout. Therefore, each $\mathcal{T}_i$ is closed under intersection and union.

We define two partitions of $E$. For $i = 1, 2$, let nonempty sets $A^i_1, A^i_2, \ldots, A^i_{k(i)}$ form a partition of $E$, defined in the following manner: Distinct $e, f \in E$ are both in $A^i_j$, $1 \leq j \leq k(i)$, if for each $T \in \mathcal{T}_i$, $\{e, f\} \subset T$ or $\{e, f\} \cap T = \emptyset$. That is, $e$ and $f$ are in the same block of the partition for matroid $M_i$, if each tight rank inequality for $z$ uses both or neither of $e$ and $f$

(it is easy to check that this is an equivalence relation, so that these blocks are well defined).

Because we assumed that $z_e > 0$ for all $e \in E$ and because

$$\mathcal{P}_{\mathcal{I}(M_i)} = \left\{ x \in \mathbf{R}_+^E \ : \ \sum_{e \in T} x_e \le r_{M_i}(T), \ \forall\, T \subset E \right\},$$

the extreme point $z$ is the *unique* solution of the equations

$$\sum_{e \in T} x_e = r_{M_1}(T), \quad \forall\, T \in \mathcal{T}_1;$$
$$\sum_{e \in T} x_e = r_{M_2}(T), \quad \forall\, T \in \mathcal{T}_2.$$

Therefore, the points

$$x(T), \quad T \in \mathcal{T}_1 \cup \mathcal{T}_2$$

span $\mathbf{R}^E$.

Notice that each $T \in \mathcal{T}_i$ is the union of some blocks $A^i_j$. Therefore, each characteristic vector $x(T)$, for a set $T \in \mathcal{T}_i$, is the sum of some characteristic vectors $x(A^i_j)$. Therefore, the points $x(A^i_j)$, $i = 1, 2$, also span $\mathbf{R}^E$. Therefore, $k(1) + k(2) \ge |E|$. In fact, $k(1) + k(2) > |E|$, as we have the linear-dependence relation $\sum_{j=1}^{k(1)} x(A^1_j) = \sum_{j=1}^{k(2)} x(A^2_j)$ (equal to the all-one vector).

Without loss of generality, we may assume that $k(1) > |E|/2$. Therefore, at least one of the $A^1_j$ must contain exactly one element. Without loss of generality, we may assume that $A^1_1 = \{f\}$.

Let

$$U := \bigcup \{T \in \mathcal{T}_1 \ : \ f \notin T\},$$

and let

$$V := \bigcap \{T \in \mathcal{T}_1 \ : \ f \in T\}.$$

Because $\mathcal{T}_1$ is closed under intersection and union, $U$ and $V$ are in $\mathcal{T}_1$.

Now, consider $e \in V \setminus U$. We have $e \in T$ if and only if $f \in T$ for every $T \in \mathcal{T}_1$. Therefore, $e$ and $f$ are in the same block $A^1_j$. However, the block containing $f$ is $A^1_1 = f$; therefore, $f = e$ and $V \setminus U = \{f\}$. Therefore, $U + f = U \cup V$, and because $U$ and $V$ are in $\mathcal{T}_1$, we have $U + f \in \mathcal{T}_1$. Hence,

$$z_f = \sum_{e \in U+f} z_e - \sum_{e \in U} z_e = r_{M_1}(U + f) - r_{M_1}(U),$$

which is either 0 or 1.

If $z_f = 0$, then let $z'$ be the projection of $z$ onto $\mathbf{R}^{E-f}$. Clearly $z' \in \mathcal{P}_{\mathcal{I}(M_1\setminus f)} \cap \mathcal{P}_{\mathcal{I}(M_2\setminus f)}$, as $r_{M_i\setminus f}(T) = r_{M_i}(T)$ for $T \in E - f$. By the inductive hypothesis, $z' \in \mathcal{P}_{\mathcal{I}(M_1\setminus f)\cap\mathcal{I}(M_2\setminus f)}$. Therefore,

$$z' = \sum_{S\in\mathcal{I}(M_1\setminus f)\cap\mathcal{I}(M_2\setminus f)} \lambda'_S x'(S),$$

where $\lambda'_S \geq 0$,

$$\sum_{S\in\mathcal{I}(M_1\setminus f)\cap\mathcal{I}(M_2\setminus f)} \lambda'_S = 1,$$

and $x'(S)$ is the characteristic vector of $S$ in $\mathbf{R}^{E-f}$. Now, let

$$\lambda_S := \begin{cases} \lambda'_S, & \text{for } S \in \mathcal{I}(M_1)\cap\mathcal{I}(M_2) \text{ such that } f \notin S \\ 0, & \text{for } S \in \mathcal{I}(M_1)\cap\mathcal{I}(M_2) \text{ such that } f \in S \end{cases}.$$

Then we have

$$z = \sum_{S\in\mathcal{I}(M_1)\cap\mathcal{I}(M_2)} \lambda_S x(S),$$

with $\lambda_S \geq 0$,

$$\sum_{S\in\mathcal{I}(M_1)\cap\mathcal{I}(M_2)} \lambda_S = 1$$

[here, $x(S)$ is the characteristic vector of $S$ in $\mathbf{R}^E$]. Therefore, $z \in \mathcal{P}_{\mathcal{I}(M_1)\cap\mathcal{I}(M_2)}$.

If $z_f = 1$, then let $z'$ be the projection of $z$ onto $\mathbf{R}^{E-f}$. Clearly, $z' \in \mathcal{P}_{\mathcal{I}(M_1/f)} \cap \mathcal{P}_{\mathcal{I}(M_2/f)}$, as $r_{M_i/f}(T) = r_{M_i}(T) - r_{M_i}(\{f\}) = r_{M_i}(T) - 1$, for $T \in E - f$. By the inductive hypothesis, $z' \in \mathcal{P}_{\mathcal{I}(M_1/f)\cap\mathcal{I}(M_2/f)}$. Therefore,

$$z' = \sum_{S\in\mathcal{I}(M_1/f)\cap\mathcal{I}(M_2/f)} \lambda'_S x'(S),$$

where $\lambda'_S \geq 0$,

$$\sum_{S\in\mathcal{I}(M_1/f)\cap\mathcal{I}(M_2/f)} \lambda'_S = 1,$$

and $x'(S)$ is the characteristic vector of $S$ in $\mathbf{R}^{E-f}$. Now, let

$$\lambda_S := \begin{cases} \lambda'_S, & \text{for } S \in \mathcal{I}(M_1)\cap\mathcal{I}(M_2) \text{ such that } f \in S \\ 0, & \text{for } S \in (\mathcal{I}(M_1)\cap\mathcal{I}(M_2) \text{ such that } f \notin S \end{cases}.$$

Then we have

$$z = \sum_{S\in\mathcal{I}(M_1)\cap\mathcal{I}(M_2)} \lambda_S x(S),$$

with $\lambda_S \geq 0$,

$$\sum_{S \in \mathcal{I}(M_1) \cap \mathcal{I}(M_2)} \lambda_S = 1$$

[here, $x(S)$ is the characteristic vector of $S$ in $\mathbf{R}^E$]. Therefore, $z \in \mathcal{P}_{\mathcal{I}(M_1) \cap \mathcal{I}(M_2)}$. ■

**Exercise (Intersection of three matroid polytopes).** Give an example of three matroids $M_i$ on the same ground set, so that $\mathcal{P}_{\mathcal{I}(M_1)} \cap \mathcal{P}_{\mathcal{I}(M_2)} \cap \mathcal{P}_{\mathcal{I}(M_3)}$ has a fractional extreme point. *Hint:* A three-element ground set will suffice.

## 3.5 Further Study

Whiteley's (1992) work contains much more information concerning the connection between matroids and statics. Recski (1988) provides connections between matroids and electrical networks as well as statics. The article by Lee and Ryan (1992) is a broader survey of algorithms and applications of matroids.

# 4

# *Matching*

Recall that a *matching* of a graph $G$ is a set $S \subset E(G)$ such that $|\delta_G(v) \cap S| \leq 1$, $\forall\, v \in V(G)$. Also, the matching $S$ is perfect if $|\delta_G(v) \cap S| = 1$, $\forall\, v \in V(G)$. We have already studied matchings in bipartite graphs in some detail. König's Theorem provides a characterization of maximum-cardinality matchings for bipartite graphs (see the bipartite matching example, pp. 85, 100, and see p. 44). The total unimodularity of the vertex-edge incidence matrix of a bipartite graph yields a characterization of the characteristic vectors of matchings in bipartite graphs as extreme points of a polytope (see p. 44). The Matroid-Intersection Algorithms provide efficient methods for finding maximum-cardinality and maximum-weight matchings in bipartite graphs (see Chapter 3). In this chapter, an efficient direct algorithm is provided for finding a maximum-weight matching in a (complete) bipartite graph.

The study of matchings in *nonbipartite* graphs is more complicated. We will study an efficient algorithm for the problem of finding a maximum-cardinality matching in a general graph. Additionally, an inequality description of the convex hull of the characteristic vectors of matchings of a general graph is provided. Finally, some applications of minimum-weight matchings are described.

## 4.1 Augmenting Paths and Matroids

Let $S$ be a matching of $G$. A path or cycle $P$ of $G$ is *alternating* with respect to $S$ if the elements of $P$ alternate, along the path or cycle, between elements of $S$ and elements of $E(G) \setminus S$. A vertex $v \in V(G)$ is *exposed* (with respect to $S$) if $\delta_G(v) \cap S = \emptyset$. Vertices that are not exposed are *covered*. An alternating path is *augmenting* if its endpoints are left exposed by $S$.

**Berge's Theorem.** *A matching $S$ of $G$ is of maximum cardinality if and only if $G$ has no augmenting path with respect to $S$.*

*Proof.* Let $P$ be an augmenting path with respect to $S$, and let $S' := S \Delta P$. $S'$ is a matching of $G$ such that $|S'| > |S|$.

Conversely, suppose that $S$ is a matching of $G$ that is not of maximum cardinality. Let $S'$ be a matching of $G$ such that $|S'| > |S|$. Consider $C := S' \Delta S$. The graph $G.C$ has maximum degree 2. Moreover, each nontrivial component has its edge set as either an alternating path or cycle (with respect to $S$). Because $|S'| > |S|$, $G.C$ must have some component with more edges from $S'$ than from $S$. Any such component is an augmenting path with respect to $S$. ■

**Theorem (Matching matroid).** *Let $G$ be an arbitrary graph, with $W \subset V(G)$. Then $M$ defined by $E(M) := W$, and*

$$\mathcal{I}(M) := \{X \subset W \ : \ G \text{ has a matching that covers } X\}.$$

*is a matroid.*

*Proof.* I1 and I2 obviously hold for $M$, so I3 is demonstrated here. Suppose that $X \in \mathcal{I}(M)$ and $Y \in \mathcal{I}(M)$ with $|Y| > |X|$. Let $S_X$ and $S_Y$ denote matchings that cover $X$ and $Y$, respectively. We may assume that all elements of $Y \setminus X$ are left uncovered by $S_X$; otherwise we would have some $v \in Y \setminus X$ with the property that the matching $S_X$ covers $X + v$, and we would be done. Now, consider $C := S_X \Delta S_Y$. As in the proof of Berge's Theorem, each nontrivial component of $G.C$ has its edge set as either an alternating path or cycle (with respect to $S_X$). Consider the vertices of $G.C$ that have degree 2. Each such vertex $v$ has the property that it is in $Y$ only if it is in $X$ (by our previous assumption). Therefore, each (alternating) cycle of $G.C$ has at least as many vertices in $X$ as in $Y$. Moreover, each (alternating) path of $G.C$ has at least as many *interior* vertices in $X$ as in $Y$. Therefore, because $|Y| > |X|$, there is some (alternating) path of $G.C$ with more endpoints in $Y$ than in $X$. Consider such a path $P$. Obviously, the endpoints of $P$ are in $V(G) \setminus (X \cap Y)$. Neither endpoint of $P$ can be in $X \setminus Y$, and at least one endpoint must be in $Y \setminus X$ for $P$ to have more vertices in $Y$ than in $X$. All vertices of $X$ that were covered by $S_X$ are covered by $S_X \Delta P$ in addition to any endpoint of $P$ that is in $Y \setminus X$ (there is at least one such endpoint). The result follows. ■

A *matching matroid* is any matroid that arises as in the statement of the theorem.

**Problem [Scheduling, continued (see p. 59)].** Recall the matroid described in the Scheduling Problem. Demonstrate that this matroid is a matching matroid.

**Problem (Mismatching matroid).** Let $G$ be an arbitrary graph, with $W \subset V(G)$. Define $M$ by $E(M) := W$, and

$$\mathcal{I}(M) := \{X \subset W : \text{all elements of } X \text{ are left exposed by some maximum-cardinality matching of } G\}.$$

Prove that $M$ is a matroid and describe a matroid-theoretic connection between matching matroids and these "mismatching matroids."

## 4.2 The Matching Polytope

We also have a characterization of the characteristic vectors of matchings as the extreme points of a polytope.

**Matching-Polytope Theorem.** *Let $G$ be a graph with no loops. The convex hull of the characteristic vectors of matchings of $G$ is the solution set of*

$$\text{(i)} \qquad -x_e \le 0, \quad \forall\, e \in E(G);$$

$$\text{(ii)} \qquad \sum_{e \in \delta_G(v)} x_e \le 1, \quad \forall\, v \in V(G);$$

$$\text{(iii)} \qquad \sum_{e \in E(G[W])} x_e \le \frac{|W|-1}{2}, \quad \forall\, W \subset V(G) \text{ with } |W| \ge 3 \text{ odd.}$$

*Proof.* Let $\mathcal{M}(G)$ denote the set of matchings of $G$. Because $G$ has no loops, the characteristic vectors of all single edges, together with the characteristic vector of the empty set, form a set of $|E(G)|+1$ affinely independent points. Therefore, the polytope $\mathcal{P}_{\mathcal{M}(G)}$ is full dimensional.

Our goal is to show that every facet of $\mathcal{P}_{\mathcal{M}(G)}$ is described by an inequality of the form *(i)*, *(ii)*, or *(iii)*. Let

$$(*) \qquad \sum_{e \in E(G)} \alpha(e) x_e \le \beta$$

describe a facet of $\mathcal{P}_{\mathcal{M}(G)}$. If matching $S$ satisfies

$$\sum_{e \in E(G)} \alpha(e) x_e(S) = \beta,$$

then $S$ is *tight* for $(*)$.

*Case 1:* $\alpha(e) < 0$ for some $e \in E(G)$.

In this case, no matching $S$ that is tight for $(*)$ contains $e$, because, for such an $e$, $x(S-e)$ would violate $(*)$. Therefore, $x_e(S)=0$ for all matchings $S$ that are tight for $(*)$, so $x(S)$ satisfies $(i)$ as an equation. Because $\mathcal{P}_{\mathcal{M}(G)}$ is full dimensional and because $(i)$ is valid for $\mathcal{P}_{\mathcal{M}(G)}$, $(*)$ must be a positive multiple of $(i)$.

*Case 2:* There is a vertex $v \in V(G)$ that is met by every matching that is tight for $(*)$.

Then

$$\sum_{e \in \delta_G(v)} x_e(S) = 1,$$

for all matchings $S$ that are tight for $(*)$, so $x(S)$ satisfies *(ii)* as an equation. Because $\mathcal{P}_{\mathcal{M}(G)}$ is full dimensional and because *(ii)* is valid for $\mathcal{P}_{\mathcal{M}(G)}$, $(*)$ must be a positive multiple of *(ii)*.

*Case 3:* $\alpha(e) \geq 0$ for all $e \in E(G)$, and for every $v \in V(G)$, there is some matching $S$ that is tight for $(*)$ that leaves $v$ exposed.

Define a graph $G_+$ by

$$E(G_+) := \{e \in E(G) \ : \ \alpha(e) > 0\},$$

and

$$V(G_+) := \{v \in V(G) \ : \ v \text{ is an endpoint of some } e \in E(G_+)\}.$$

We analyze Case 3 by means of a series of claims.

*Claim 1:* $G_+$ is connected.

If $G_+$ were the disjoint union of nonempty $G_1$ and $G_2$, then for $i=1,2$, let

$$\alpha^i(e) := \begin{cases} \alpha(e) & \text{if } e \in E(G_i) \\ 0 & \text{if } e \in E(G) \setminus E(G_i) \end{cases}.$$

Then $\alpha(e) = \alpha^1(e) + \alpha^2(e)$, for all $e \in E(G)$. For $i=1,2$, let $S^i$ be a matching that maximizes

$$\sum_{e \in E(G)} \alpha^i(e) x_e(S^i),$$

and let $\beta^i$ be the optimal value. We can assume that $S^i \subset E(G_i)$. Then

$$\sum_{e \in E(G)} \alpha^i(e) x_e \leq \beta^i$$

is valid for $\mathcal{P}_{\mathcal{M}(G)}$, for $i = 1, 2$. Moreover, $(*)$ is the sum of these two inequalities, which contradicts the assumption that $(*)$ describes a facet.

*Claim 2:* If $S \in \mathcal{M}(G)$ is tight for $(*)$, then $S$ leaves at most one element of $V(G_+)$ exposed.

Suppose that there is a matching that is tight for $(*)$ and leaves a pair of elements of $V(G_+)$ exposed. Among all such matchings and their exposed pairs, choose a matching $S$ and an exposed pair $u, v$ so that $u$ and $v$ are the minimum distance apart in $G_+$. Let $P$ be a shortest path connecting $u$ and $v$ in $G_+$. Clearly $P$ cannot consist of a single edge, say $e$. If that were the case, then $S + e$ would be a matching that violates $(*)$, as $S$ is tight for $(*)$ and $\alpha(e) > 0$. Therefore, we can choose a vertex $w$ on $P$ that is distinct from $u$ and $v$. The vertex $w$ is met by $S$ due to the choice of the pair $u, v$.

Let $S' \in \mathcal{M}(G)$ be a matching that is tight for $(*)$ and leaves $w$ exposed (such an $S'$ exists by the hypothesis of Case 3). Then $S \Delta S'$ contains an alternating path $Q$ that has $w$ as an endpoint. Because $S$ and $S'$ are both tight for $(*)$, we have

$$\sum_{e \in S} \alpha(e) + \sum_{e \in S'} \alpha(e) = 2\beta.$$

Now, $S \Delta Q = (S \setminus Q) \cup (S' \cap Q)$, and $S' \Delta Q = (S' \setminus Q) \cup (S \cap Q)$. Therefore, we have

$$\sum_{e \in S \Delta Q} \alpha(e) + \sum_{e \in S' \Delta Q} \alpha(e) = 2\beta.$$

Because $S \Delta Q$ and $S' \Delta Q$ are matchings, it must be that

$$\sum_{e \in S \Delta Q} \alpha(e) \le \beta,$$

and

$$\sum_{e \in S' \Delta Q} \alpha(e) \le \beta.$$

Therefore, we can conclude that the matching $S \Delta Q$ is tight for $(*)$. However, $S \Delta Q$ leaves $w$ exposed, as well as at least one of $u$ and $v$ (remember, $u$ and $v$ are left exposed by $S$ and $Q$ is an alternating path that meets $w$; so $Q$ can meet at most one of $u$ and $v$ and only as an endpoint). This contradicts the choice of $S$ and the pair $u, v$.

*Claim 3:* For every $v \in V(G_+)$, the graph obtained when $v$ is deleted (along with its incident edges) from $G_+$ has a perfect matching.

By the hypothesis of Case 3, there is a matching that is tight for $(*)$ and leaves exposed $v \in V(G_+)$. Choose such a matching $S$ so that $S \subset E(G_+)$ [just delete edges not in $E(G_+)$]. By Claim 2, $S$ leaves no vertex of $G_+$ exposed besides $v$. Therefore, $S$ is a perfect matching of the graph obtained from $G_+$ by deleting $v$.

*Claim 4:* Let $W := V(G_+)$. Every matching that is tight for $(*)$ contains exactly $(|W| - 1)/2$ edges of $G[W]$.

Let $S$ be a matching that is tight for $(*)$. As in Claim 3, we can assume that $S$ is contained in $E(G[W])$. By Claim 2, $S$ leaves at most one element of $W$ exposed. Therefore, $S$ contains at least $(|W| - 1)/2$ edges of $G[W]$. Claim 3 implies that $|W|$ is odd; therefore, $|S| \le (|W| - 1)/2$. Therefore, $|S| = (|W| - 1)/2$, and $x(S)$ satisfies *(iii)* as an equation.

Because $\mathcal{P}_{\mathcal{M}(G)}$ is full dimensional and because *(iii)* is valid for $\mathcal{P}_{\mathcal{M}(G)}$, $(*)$ must be a positive multiple of *(iii)*. ∎

**Exercise (Weighted matching).** Consider the following "envelope graph" $G$ with edge weights $c(e)$ as indicated and the associated *linear program* $\max \sum_{e \in E(G)} c(e) x_e$ subject to *(i)*, *(ii)*, and $x_e \in \mathbf{Z}, \forall\, e \in E(G)$.

a. Convince yourself that the optimal solution of this integer program has objective value 21.
b. Prove that the optimal objective value of the associated linear-programming relaxation is 30 by displaying feasible solutions to it and its dual having objective value 30.
c. Next, include constraints *(iii)* as well, and prove that the optimal objective value of the linear-programming relaxation is 21 by displaying a feasible solution to its dual having objective value 21.

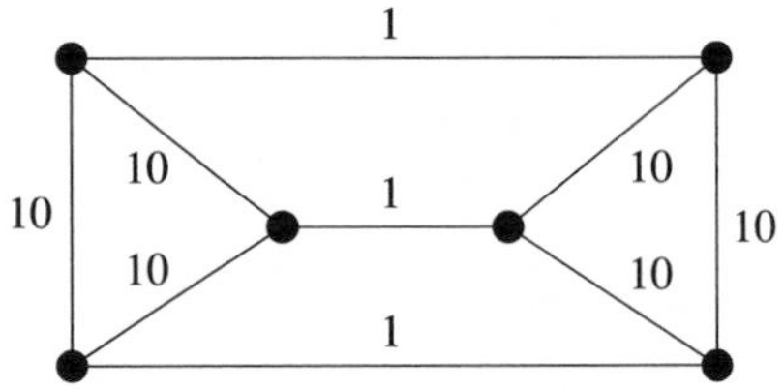


**Problem (Facets of a matching polytope).** Let $G$ be a complete graph on at least four vertices. Show that the inequalities *(i)*, *(ii)*, and *(iii)* describe facets of $\mathcal{P}_{\mathcal{M}(G)}$.

## 4.3 Duality and a Maximum-Cardinality Matching Algorithm

We also have a generalization of König's Theorem. An *odd-set cover* of a graph $G$ is a set $\mathcal{W} = (\{W_1, W_2, \ldots, W_k\}; \{v_1, v_2, \ldots, v_r\})$, where $v_j \in V(G)$, each $W_i \subset V(G)$ has odd cardinality ($\geq 3$), and every edge of $G$ has a $v_j$ as an endpoint or has both endpoints in the same $W_i$. The *capacity* of the odd-set cover $\mathcal{W}$ is

$$r + \sum_{i=1}^{k} \frac{|W_i| - 1}{2}.$$

The idea of an odd-set cover can be motivated by the inequality description of $\mathcal{P}_{\mathcal{M}(G)}$ and linear-programming duality. The maximum cardinality of a matching of $G$ is equal to the maximum value of $\sum_{e \in E(G)} x_e$ subject to inequalities *(i)–(iii)*. The dual of this linear program is

$$\min \sum_{v \in V(G)} y_v + \sum_{\substack{W \subset V(G)\,:\\ |W| \geq 3,\ \text{odd}}} \frac{|W| - 1}{2} \pi_W$$

subject to:

$$y_{v_1} + y_{v_2} + \sum_{\substack{W \subset V(G)\,:\, e \subset W\\ |W| \geq 3,\ \text{odd}}} \pi_W \geq 1, \qquad \forall\ e = \{v_1, v_2\} \in E(G);$$

$$y_v \geq 0, \qquad \forall\ v \in V(G);$$

$$\pi_W \geq 0, \qquad \forall\ W \subset V(G)\ :\ |W| \geq 3, \text{odd}.$$

The characteristic vector of an odd-set cover is a feasible solution to this dual linear program, and the objective value of the solution is the capacity of the cover. Therefore, the capacity of an odd-set cover is an upper bound on the cardinality of a matching. In fact, we demonstrate the following stronger result.

**Matching Duality Theorem.** *The maximum cardinality of a matching of a loop-free graph $G$ is equal to the minimum capacity of an odd-set cover of $G$.*

**Problem (Disjoint odd-set cover).** Let $\mathcal{W} = (\{W_1, W_2, \ldots, W_k\};\ \{v_1, v_2, \ldots, v_r\})$ be an *arbitrary* minimum-capacity odd-set cover of a simple graph $G$. Describe an efficient procedure that alters $\mathcal{W}$ to create a new minimum-capacity odd-set cover $\mathcal{W}' = \left(\{W'_1, W'_2, \ldots, W'_{k'}\};\ \{v'_1, v'_2, \ldots, v'_{r'}\}\right)$ such that each $v'_j$ is not an element of each $W'_i$, and the $W'_i$ are disjoint from one another. (*Note*: Be sure to explain why your procedure terminates, why your $\mathcal{W}'$ is an odd-set cover, and why your $\mathcal{W}'$ has minimum capacity.)

**Problem (Tutte's Perfect-Matching Theorem).** Let $G$ be a simple graph. For $W \subset V(G)$, let $\kappa_{\text{odd}}(G[V(G) \setminus W])$ denote the number of components with an odd number of vertices in the subgraph of $G$ induced by $V(G) \setminus W$. Note that an isolated vertex is an "odd component." Use the Matching Duality Theorem to prove that $G$ has a perfect matching if and only if $\kappa_{\text{odd}}(G[V(G) \setminus W]) \leq |W|$ for all $W \subset V(G)$. *Hint:* Use the fact that you can choose a minimum-capacity odd-set cover to be "disjoint."

The proof of the Matching Duality Theorem follows from Edmonds's Maximum-Cardinality Matching Algorithm. Edmonds's algorithm is based on the following result.

**Shrinking Lemma.** *Let $G$ be an undirected graph, and let $S$ be a matching of $G$. Let $C$ be a cycle with $|C| = 2l + 1$ for some positive integer $l$. Suppose that $|S \cap C| = l$ and $S \setminus C$ is vertex disjoint from $C$. Construct a graph $G'$ by shrinking $C$ to a single vertex. Then $S' := S \setminus C$ is a maximum-cardinality matching of $G'$ if and only if $S$ is a maximum-cardinality matching of $G$.*

*Proof.* Suppose that $S$ is not a maximum-cardinality matching of $G$. Let $P$ be an augmenting path with respect to $S$ (in $G$). If $P$ is vertex disjoint from $C$, then $P$ is also augmenting with respect to $S'$ (in $G'$). So suppose that $P$ is not vertex disjoint from $C$. At least one endpoint of $P$, say $v$, is not on $C$, as only one vertex of $C$ is exposed, but both endpoints of $P$ are exposed. Let $w$ be the first vertex of $C$ encountered while traversing $P$ starting at $v$. Then the subpath $P'$ of $P$ that extends from $v$ to $w$ is augmenting with respect to $S'$ (in $G'$). Thus $S'$ is not a maximum-cardinality matching of $G'$.

Conversely, suppose that $S'$ is not a maximum-cardinality matching of $G'$. Let $T'$ be a matching of $G'$ with $|T'| > |S'|$. Now, expand $C$ to recover $G$. Then, $T'$ is a matching of $G$ that covers at most one vertex of $C$. We can choose $l$ elements of $C$ to adjoin to $T'$ to get a matching $T$ of $G$. Because $|T| = |T'| + l > |S'| + l = |S|$, $S$ is not a maximum-cardinality matching of $G$. ■

The algorithm uses the idea of an "alternating forest" to find augmenting paths. An *alternating forest* with respect to a matching $S$ of $G$ is a subgraph $H$ such that

1. $E(H)$ is a forest;
2. each component of $H$ contains exactly one exposed vertex, called the *root*, and every exposed vertex is the root of a component of $H$;

3. vertices of $H$ are called *odd* or *even* depending on their distance to their root, and each odd vertex has degree 2 in $H$ and one of the two incident edges in $H$ is in $S$.

This sounds more complicated than it really is. A picture clarifies the situation. The following picture shows what a component of $H$ might look like. Wavy edges are matching edges. A component could consist of an isolated root. Note that every matching edge that is not in $H$ is vertex disjoint from $H$. Also, every vertex that is not in $H$ is covered.

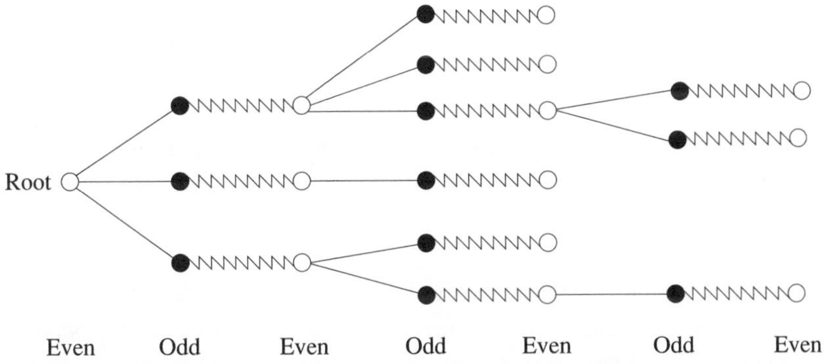


**Edmonds's Maximum-Cardinality Matching Algorithm**

Let $S_k$ be a matching of $G$ of cardinality $k$. (Can take $k = 0$).

0. (Seed). Seed a forest $H$ with just the exposed vertices as isolated roots with no edges. Let $G' \leftarrow G$, $S' \leftarrow S_k$. Next, repeatedly apply any of Steps 1–3, as they are applicable.
1. (Grow). If there is an edge $e \in E(G') \setminus E(H)$ with an even endpoint $x$ and the other endpoint $y \notin V(H)$, then $y$ is met by some $f \in S' \setminus E(H)$. Moreover, the other endpoint $z$ of $f$ is not in $V(H)$, so we can redefine $H$ by letting $E(H) \leftarrow E(H) + e + f$, $V(H) \leftarrow V(H) + y + z$.
2. (Augment). If there is an edge $e \in E(G') \setminus E(H)$ with its endpoints being even vertices of *different* components of $H$, then $E(H) + e$ contains a path $P$ between the roots of the components containing the endpoints of $e$. The path $P$ is augmenting with respect to $S'$, so we can let $S' \leftarrow S' \Delta P$ to increase the cardinality of the matching of $G'$. By repeatedly unshrinking all shrunken cycles, we recover the original graph $G$ and, by applying the Shrinking Lemma (repeatedly), a matching $S_{k+1}$ of $G$ such that $|S_{k+1}| = k + 1$. Let $k \leftarrow k + 1$, and go to Step 0.

3. (Shrink). If there is an edge $e \in E(G') \setminus E(H)$ with its endpoints being even vertices of the *same* component of $H$, let $P$ be the path from either endpoint of $e$ to the root of the component. Let $S' \leftarrow S' \Delta P$ (note that this does not alter the cardinality of $S'$), and shrink the unique cycle in $E(H) + e$ (note that we switched $S'$ to satisfy the conditions of the Shrinking Lemma). The shrunken cycle becomes the root of its component in the alternating forest.
4. (Optimality). If none of Steps 1–3 is applicable, then $S_k$ is a maximum-cardinality matching of $G$.

We will be satisfied with a crude bound on the number of steps required by Edmonds's algorithm.

**Proposition (Finiteness and efficiency of Edmonds's Cardinality Matching Algorithm).** *Edmonds's Maximum-Cardinality Matching Algorithm terminates in $\mathcal{O}(|V(G)|^4)$ time.*

*Proof.* First we note that the number of augmentations is bounded by $|V(G)|/2$ (consider how many edges a matching could have). Let us consider the course of the algorithm between augmentations. The number of growth steps is bounded by $|V(G)|/2$ (consider how many edges a forest could have). The number of shrink steps is bounded by $|V(G)|/2$ (consider how many vertices are "lost" when we shrink). Therefore, the total number of iterations of Steps 1, 2, and 3 is $\mathcal{O}|V(G)|^2$. Using the simplest of data structures, we can easily carry out each step in $\mathcal{O}|V(G)|^2$ time, so the total running time in terms of elementary operations is $\mathcal{O}(|V(G)|^4)$. ■

**Lemma (Maximum-cardinality matching in a shrunken graph).** *At the conclusion of Edmonds's Maximum-Cardinality Matching Algorithm, $S'$ is a maximum-cardinality matching of $G'$.*

*Proof.* Let $H$ be the forest of $G'$ at the conclusion of the algorithm. Let $\mathcal{E}$ be the set of even vertices of $H$, and let $\mathcal{O}$ be the set of odd vertices of $H$. Let $\mathcal{U}$ be the set of vertices of $G'$ that are not in $H$. All vertices of $H$ are covered except for the roots. Moreover, because of the alternating nature of the forest, all non-root vertices of $H$ are met by elements of $S' \cap E(H)$. Therefore, no matching edge extends between $H$ and $\mathcal{U}$. However, all elements of $\mathcal{U}$ are covered by $S'$ (otherwise they would be roots); therefore, $S' \cap E(G'[\mathcal{U}])$ is a perfect matching of $G'[\mathcal{U}]$, and $|S' \cap E(G'[\mathcal{U}])| = |\mathcal{U}|/2$. Moreover, by the alternating structure of $H$, we have $|S' \cap E(H)| = |\mathcal{O}|$.

If $|\mathcal{U}| > 2$, then choose $v \in \mathcal{U}$ and let $\mathcal{W} := (\{\mathcal{U} - v\}; \mathcal{O} + v)$.

We claim that $\mathcal{W}$ is an odd-set cover of $G'$. This follows if we note that (1) every edge of $H$ is met by an element of $\mathcal{O}$, (2) every edge of $G'[U]$ either has both endpoints in $\mathcal{U} - v$, or is met by $v$, and (3) the only edges that are not in $E(H) \cup E(G'[U])$ have an endpoint in $\mathcal{O}$ (because otherwise we could grow, shrink or augment).

If, instead, we have $|\mathcal{U}| = 2$, then we modify the construction of the odd-set cover so that $\mathcal{W} := (\emptyset; \mathcal{O} + v)$. If, instead, we have $\mathcal{U} = \emptyset$, then we let $\mathcal{W} := (\emptyset; \mathcal{O})$.

In any case, it is easy to check that the capacity of $\mathcal{W}$ and the cardinality of $S'$ are both $|\mathcal{O}| + |\mathcal{U}|/2$. ∎

This lemma, together with the Shrinking Lemma, establishes the validity of Edmonds's Maximum-Cardinality Matching Algorithm.

**Theorem (Correctness of Edmonds's Cardinality Matching Algorithm).** *At the conclusion of Edmonds's Maximum-Cardinality Matching Algorithm, $S_k$ is a maximum-cardinality matching of $G$.*

**Problem (Algorithmic proof of König's Theorem).** Adjust the proof of the previous lemma (Maximum-cardinality matching in a shrunken graph) to prove König's Theorem (see p. 44).

Next, we work through an example to illustrate the algorithm.

**Example (Matching).** Consider the following graph (the matching is indicated by the wavy edges):

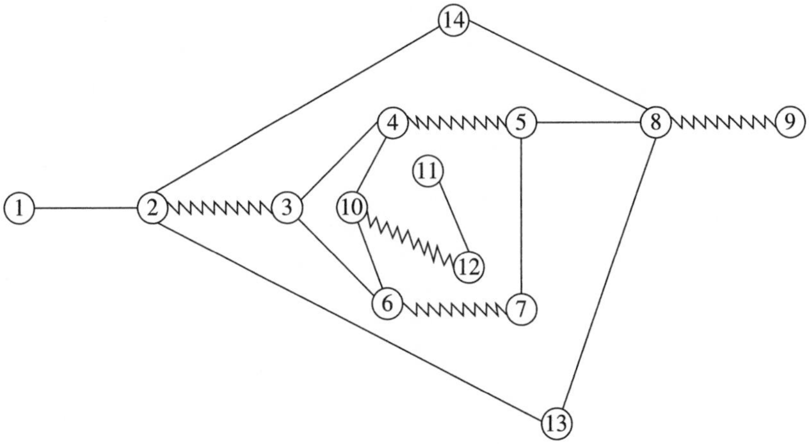

First, we repeatedly apply Step 1 of the algorithm and grow an alternating forest:

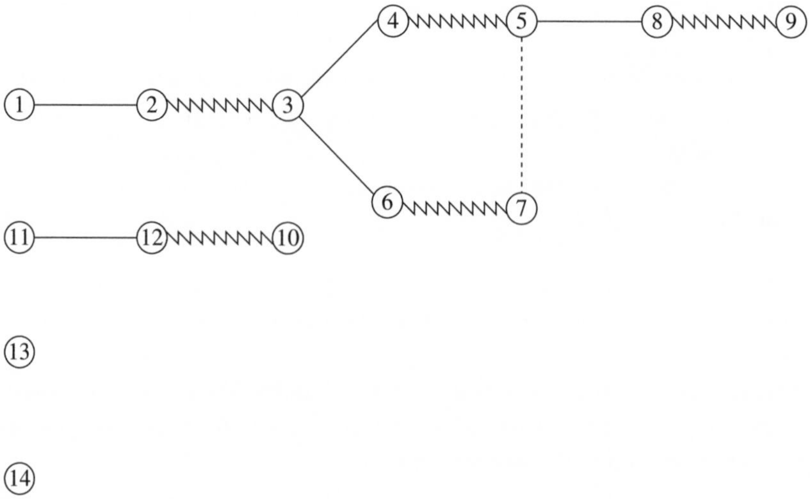


Vertices 5 and 7 are even vertices that are in the same component of the forest, and they are connected by an edge. Therefore, we can apply Step 3 of the algorithm. We alternate the matching on the path having imputed vertex sequence $1, 2, 3, 4, 5$:

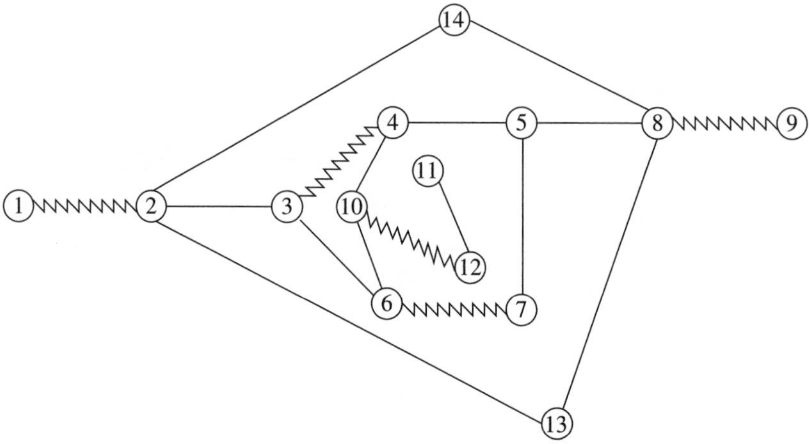

Then we shrink the cycle having imputed vertex sequence 3, 4, 5, 6, 7, 3 to obtain the graph

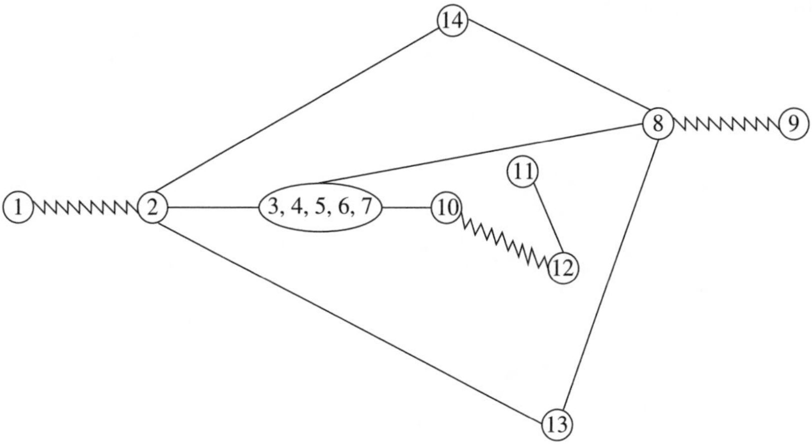


and the associated alternating forest

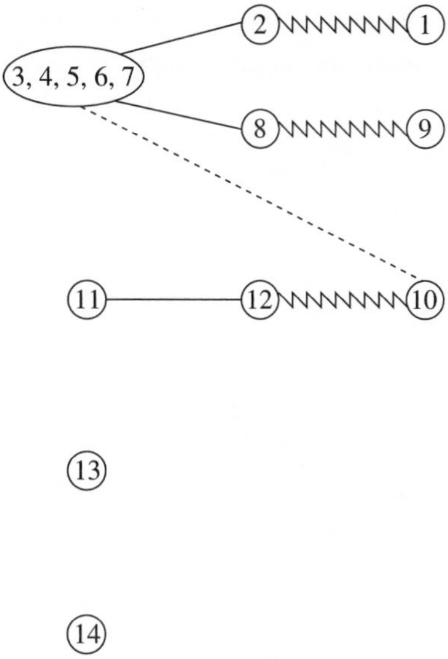


(note that the shrunken cycle becomes the root of its component). The two vertices (3, 4, 5, 6, 7) and 10 are even vertices that are in different components of the forest, and they are connected by an edge. Therefore, we can apply Step 2 of the algorithm, and we obtain the augmenting path with imputed vertex sequence (3, 4, 5, 6, 7), 10, 12, 11. Carrying out the augmentation, we obtain the following matching:

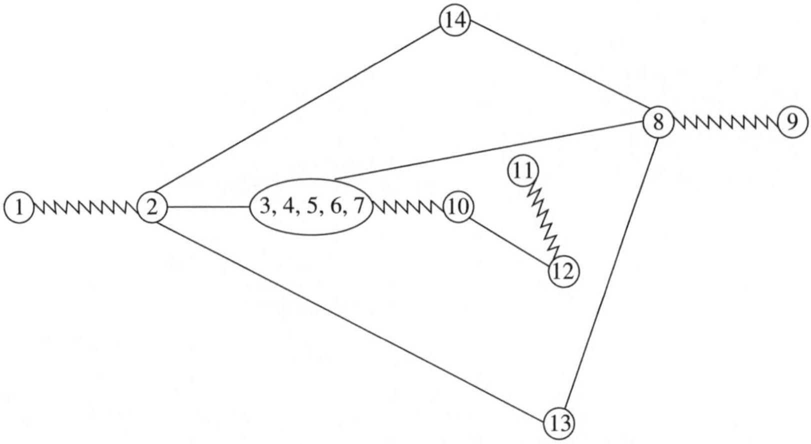


Next, we unshrink the shrunken cycle, by applying the Shrinking Lemma, to obtain a larger matching of the original graph. Vertex 10 can be matched to either vertex 4 or 6; we arbitrarily pick vertex 4. Then we can choose any perfect matching of the remaining vertices of the cycle to produce a larger matching of the original graph:

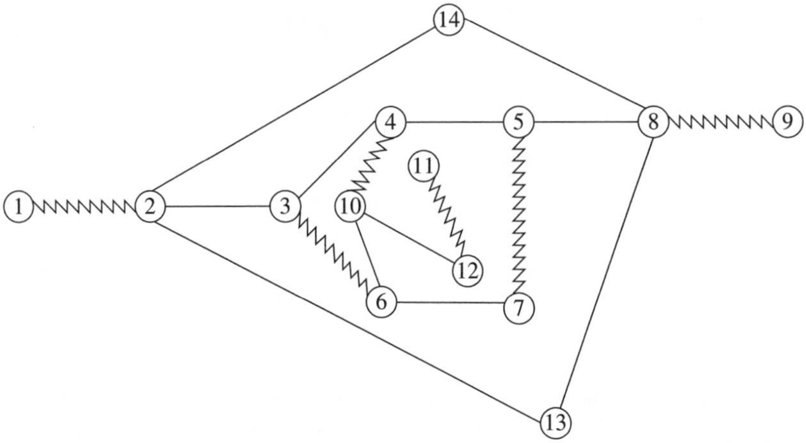

Next, we reseed, and grow the forest

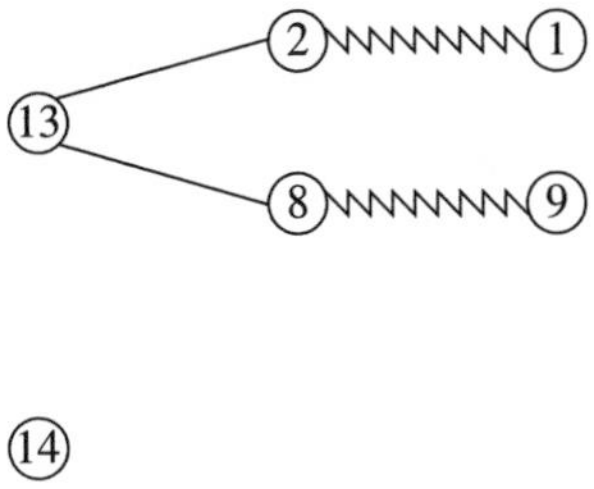


None of Steps 1–3 applies, so Step 4 of the algorithm indicates that the matching of cardinality 6 has maximum cardinality. An odd-set cover of capacity 6 is

$$\mathcal{W} := (\{\{4, 5, 6, 7, 10, 11, 12\}\}\,; \{2, 3, 8\})\,.$$ ♠

**Problem (Matching duality theorem).** The proof of validity of Edmonds's Maximum-Cardinality Matching Algorithm uses a construction of a minimum-capacity odd-set cover of the *shrunken* graph $G'$. Prove the Matching Duality Theorem by describing how to construct a minimum-capacity odd-set cover of the *original* graph $G$.

## 4.4 Kuhn's Algorithm for the Assignment Problem

Let $G$ be a complete bipartite graph, with vertex bipartition $V_1(G)$, $V_2(G)$ satisfying $n := |V_1(G)| = |V_2(G)|$. Let $c$ be a weight function on $E(G)$. The assignment problem is to find a maximum-weight perfect matching of the complete bipartite graph $G$. The assignment problem is often used as a model for assigning personnel to tasks.

The more general problem in which we do not assume that $G$ is complete is easily handled. We simply include any missing edges into $G$ and apply a very large negative weight.

The algorithm that is presented for the assignment problem is driven by considerations involving linear-programming duality. The most natural formulation of the assignment problem as an integer linear program is

$$\max \sum_{e\in E} c(e)x_e$$

subject to:

$$\sum_{e\in\delta_G(v)} x_e = 1, \qquad \forall\, v \in V(G);$$

$$x_e \geq 0, \qquad \forall\, e \in E(G).$$

Of course, the variables should be required to be integer, but we get that for free because the constraint matrix is totally unimodular and the constraint right-hand sides are integers.

The linear programming dual is

$$\min \sum_{v \in V(G)} y_v$$

subject to:

$$y_v + y_w \geq c(\{v, w\}), \qquad \forall\, e = \{v, w\} \in E(G).$$

For any choice of $y \in \mathbf{R}^{V(G)}$, we define the *transformed weight function* $\overline{c}$ by letting $\overline{c}(\{v, w\}) := c(\{v, w\}) - y_v - y_w$, for all edges $\{v, w\} \in E(G)$. Note that dual feasibility of $y$ is equivalent to nonpositivity of the transformed weight function. It is an easy observation that, for any perfect matching $F$ of $G$, $\overline{c}(F) = c(F) - \sum_{v \in V(G)} y_v$. Therefore, $F$ is a maximum-weight perfect matching with respect to $\overline{c}$ if and only if $F$ is a maximum-weight perfect matching with respect to $c$.

At all times, the algorithm maintains a dual-feasible solution $y$. A dual feasible solution can be used in a sufficient optimality criterion. We define the *equality subgraph* $G_=$ by $V(G_=) := V(G)$ and $E(G_=) := \{e \in E(G) \; : \; \overline{c}(e) = 0\}$. If $y$ is dual feasible and $F$ is any set of edges, then obviously $\overline{c}(F) \leq 0$. If $y$ is dual feasible, then any perfect matching of $G_=$ is a maximum-weight perfect matching of $G$. It is simple enough to construct an initial dual feasible solution by taking the $y_v$ to be large enough.

Therefore, the algorithm starts with a dual feasible solution $y$ and constructs a maximum-cardinality matching $F$ of $G_=$. If $F$ is perfect, then we are done. If $F$ is not perfect, we grow a maximal alternating forest $H$ with respect to $F$, using exposed $v \in V_1(G)$ as roots. At the conclusion of this phase, once $H$ is maximal, all exposed $w \in V_2(G)$ are not in the alternating forest (because otherwise we would have discovered an augmenting path with respect to $F$, contradicting the maximum cardinality of $F$).

Next, we define

$$\Delta := \max \{-\overline{c}(\{v, w\}) \; : \; v \in V_1(G) \cap V(H), \; w \in V_2(G) \setminus V(H)\}.$$

Notice that $\Delta > 0$ because otherwise we could continue to grow the forest $H$.

Finally, we update the dual solution $y$ as

$$y_v := \begin{cases} y_v - \Delta, & \text{if } v \in V_1(G) \cap V(H) \\ y_v + \Delta, & \text{if } v \in V_2(G) \cap V(H), \\ y_v, & \text{if } v \notin V(H) \end{cases}$$

and repeat (i.e., form the equality subgraph $G_=$, find a maximum-cardinality perfect matching of $G_=$, consider a maximal alternating forest, etc.) until the algorithm terminates.

The only possible termination is with a perfect matching of $G_=$ (which is a maximum-weight perfect matching of $G$). Therefore, we should be satisfied if we can establish (1) that the transformed weights remain nonnegative after a change in the dual solution $y$, and (2) that the algorithm must terminate in a reasonable number of steps.

For (1), we note that the only edges $\{v, w\}$ for which the transformed weight increases are those that have $v \in V_1(G) \cap V(H)$ and $w \in V_2(G) \setminus V(H)$. All such edges have their transformed weight increase by exactly $\Delta$, and $\Delta$ is chosen to make sure that the least negative of these is increased to 0 (so all of them will remain nonnegative).

For (2), we make two observations: (a) the number of times the cardinality of the matching can increase is just $n$; (b) between increases in the cardinality of the matching, on a dual change, the previously maximal alternating forest $H$ can grow further, and that can happen at most $n$ times. To expand on observation (b), we note that, after a dual change, all of the edges of the previous $H$ are still in $G_=$, and any edge $\{v, w\}$ that enters $G_=$ (and there is at least one) can be appended to $F$; if $w$ is exposed, then we have an augmenting path that leads to a matching of greater cardinality; if $w$ is not exposed then we also append the matching edge that touches $w$ to $F$.

It is easy to see then that the dual solution $y$ is changed at most $n^2$ times and that the number of basic computational steps between each dual change is $\mathcal{O}(n^2)$. Therefore, the total running time is $\mathcal{O}(n^4)$.

Although we make no use of it, we note that each component of $H$ has one more vertex from $V_1(G)$ than $V_2(G)$. Therefore, $|V_1(G) \cap V(H)| > |V_2(G) \cap V(H)|$ [whenever $V_1(G)$ has elements left exposed by the matching $F$ – or, equivalently, when $F$ is not perfect]. Therefore, it is easy to see that $\sum_{v \in V(G)} y_v$ decreases [by $\Delta(|V_1(G) \cap V(H)| - |V_2(G) \cap V(H)|]$ at each step. This observation can be used to produce a somewhat simpler proof that the algorithm terminates (although we do not get the polynomial bound on the running time with such a proof).

**Example (Kuhn's Assignment Algorithm).** We are in the process of solving a maximum-weight assignment problem for a problem with $V_1(G) := \{1, 2, 3, 4\}$ and $V_2(G) := \{a, b, c, d\}$. The matrix of edge weights is

$$\begin{array}{c} \\ 1 \\ 2 \\ 3 \\ 4 \end{array} \begin{array}{c} \begin{array}{cccc} a & b & c & d \end{array} \\ \begin{pmatrix} 6 & 15 & 12 & 13 \\ 18 & 8 & 14 & 15 \\ 13 & 12 & 17 & 11 \\ 18 & 16 & 14 & 10 \end{pmatrix} \end{array}$$

That is, the entry in row $v$ and column $w$ is $c(\{v, w\})$. Using $(y_1, y_2, y_3, y_4) = (15, 18, 17, 18)$ (for the row vertices) and $(y_a, y_b, y_c, y_d) = (0, 0, 0, -2)$ (for the column vertices), we compute the matrix of transformed edge weights:

$$\begin{array}{c} \\ 1 \\ 2 \\ 3 \\ 4 \end{array} \begin{array}{c} \begin{array}{cccc} a & b & c & d \end{array} \\ \begin{pmatrix} -9 & 0 & -3 & 0 \\ 0 & -10 & -4 & -1 \\ -4 & -5 & 0 & -4 \\ 0 & -2 & -4 & -6 \end{pmatrix} \end{array}$$

The equality subgraph $G_=$ is

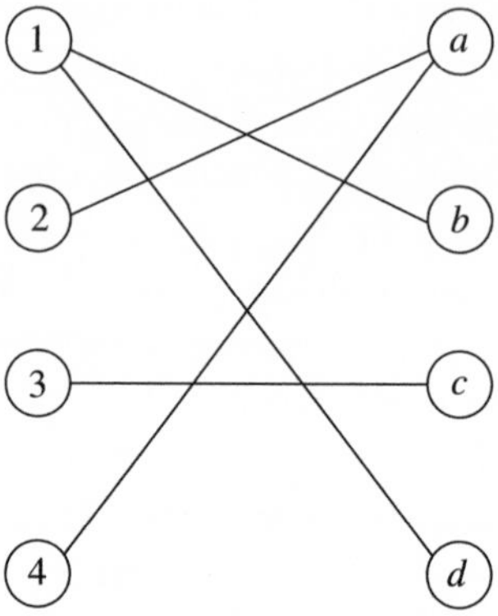


It has a maximum-cardinality matching $F$, indicated by

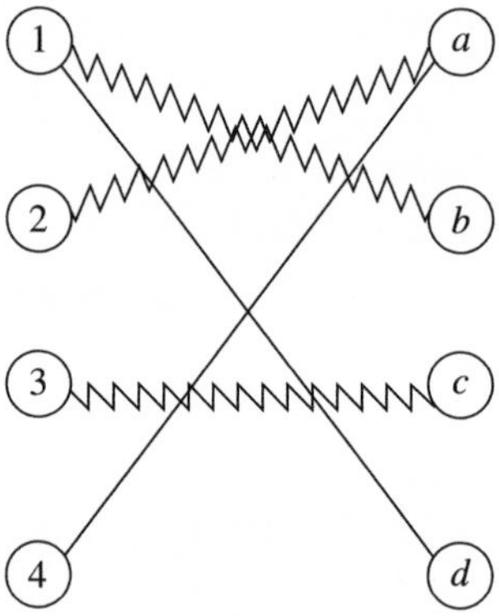

At this point a maximal alternating forest (seeded from vertex 1, the only exposed vertex in $V_1$) is $H$, indicated by

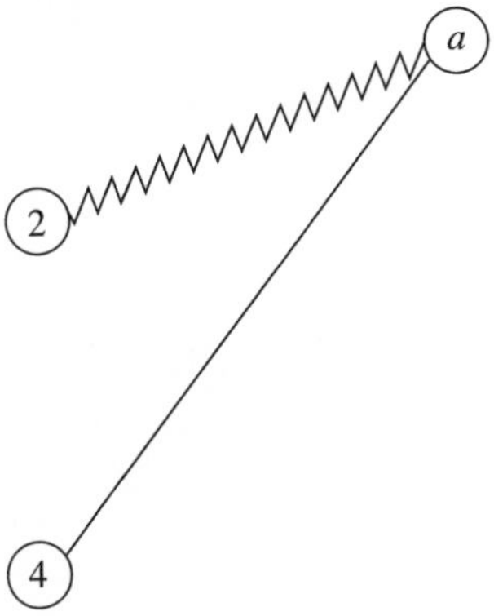


It is easy to verify that $F$ is of maximum cardinality in $G_=$, as $(V_1(G_=) \setminus V(H)) \cup (V_2(G) \cap V(H)) = \{1, 3, a\}$ is a vertex cover of $G_=$ having cardinality equal to that of $F$.

From the alternating forest $H$, we calculate $\Delta = 1$, and we update the dual variables to $(y_1, y_2, y_3, y_4) = (15, 17, 17, 17)$, $(y_a, y_b, y_c, y_d) = (1, 0, 0, -2)$ and the transformed edge weights to

$$\begin{array}{c} \\ 1 \\ 2 \\ 3 \\ 4 \end{array} \begin{array}{c} \begin{array}{cccc} a & b & c & d \end{array} \\ \begin{pmatrix} -10 & 0 & -3 & 0 \\ 0 & -9 & -3 & 0 \\ -5 & -5 & 0 & -4 \\ 0 & -1 & -3 & -5 \end{pmatrix} \end{array}$$

Edge $\{2, d\}$ enters the equality subgraph $H$ (and, by design, no edges of $H$ leave), which enables us to further grow the alternating forest to

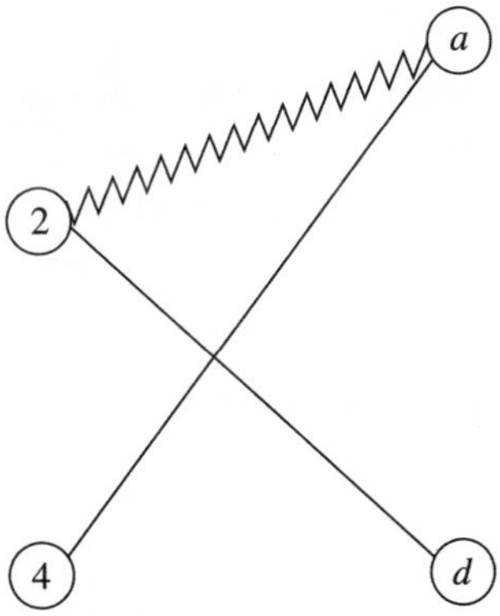

which contains (is!) an augmenting path. This leads to the matching

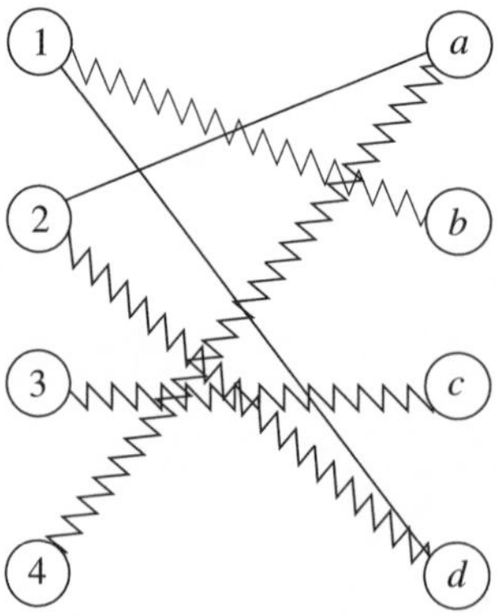


of $G_=$. This matching is perfect in $G_=$ and is thus a maximum-weight perfect matching of $G$. ♠

## 4.5 Applications of Weighted Matching

Although an algorithm for finding optimum-weight matchings in general graphs is not presented, we do discuss some applications of such an algorithm. One simple application is a method for finding a minimum-weight even path in an undirected graph with nonnegative edge weights.

**Problem (Minimum-weight even path).** Let $G$ be an undirected graph with nonnegative edge weights and a pair of distinct vertices $v$ and $w$. The goal is to develop an algorithm to find, among all $v$–$w$ paths having an even number of edges, a path having minimum weight.

Consider the following construction. Let $H := G[V(G) - v]$, and let $H'$ be a copy of $G[V(G) - w]$, where $V(H') := \{u' \ : \ u \in V(G) - w\}$. Let $M$ be a set of disjoint edges connecting each element $u$ of $V(H) - w$ with its "copy" $u'$ in $V(H') - v$. Now, we form a new graph $G'$ having $V(G') := V(H) \cup V(H')$, and $E(G') := E(H) \cup E(H') \cup M$. Each edge of $G'$ in $E(H) \cup E(H')$ gets weight equal to the weight of the corresponding edge in $G$. Edges of $G'$ in $M$ get weight 0.

a. Prove that a minimum-weight perfect matching of $G'$ can be used to find, among all $v$–$w$ paths having an even number of edges, a path having minimum weight.
b. What goes wrong if there are edges with negative weight in $G$?

Further applications of weighted matching are best explained in the context of a certain generalization of perfect matchings. For a graph $G$, let $T$ be an even-cardinality subset of $V(G)$. A subset $F$ of $E(G)$ is a *$T$-join* of $G$ if

$$|\delta_G(v) \cap F| \text{ is } \begin{cases} \text{odd}, & \text{if } v \in T \\ \text{even}, & \text{if } v \in V(G) \setminus T \end{cases}.$$

Consider the vertex-edge incidence matrix $A(G)$, the characteristic vector $x(F)$ of a set $F \subset E(G)$, and the characteristic vector $x(T)$ of an even-cardinality subset $T \subset E(G)$. It is easy to see that $F$ is a $T$-join if and only if $A(G)x(S) = x(T)$, where we do the arithmetic in **GF(2)**.

A $T$-join is *minimal* if it does not properly contain a $T$-join. Certainly every minimal $T$-join is a forest because we can repeatedly remove cycles from a $T$-join that is not a forest. We are especially interested in minimal $T$-joins because, for positive weight functions on $E(G)$, minimum-weight $T$-joins are minimal $T$-joins. In addition, for nonnegative weight functions, every minimum-weight $T$-join is the (edge) disjoint union of a minimal $T$-join and a subgraph of weight zero edges consisting of an $\emptyset$-join.

The set of $V(G)$-joins that have cardinality $|V(G)|/2$ is precisely the set of perfect matchings of $G$. By the addition of a large positive constant to each edge weight, a minimum-weight $V(G)$-join for the new weights will be a minimum-weight perfect matching with respect to the original weights. In this way, $T$-joins generalize perfect matchings.

Another motivating example comes from the problem of finding a minimum-weight $v$–$w$ path in an undirected graph $G$. Let $d$ be a weight function on $E(G)$. If the weight function $d$ is nonnegative, then we can just replace each edge of $G$ with an oppositely directed pair to form a digraph $H$. That is, edge $\{i, j\}$ of $G$ gives rise to an oppositely directed pair of edges $e'$ and $e''$ in $H$ [i.e., with $t(e') := h(e'') := i$ and $h(e') := t(e'') := j$]. We define a weight function $c$ on $E(H)$ by letting $c(e') := c(e'') := d(e)$. Then, with respect to $c$, we find a minimum-weight $v$–$w$ dipath in $H$. This procedure fails if $G$ has negative-weight edges, as $H$ would then have negative-weight dicycles. However, if we let $T := \{v, w\}$, then the set of minimal $T$-joins is precisely the set of (undirected) $v$–$w$ paths of $G$. Then, as long as $G$ contains no negative-weight cycles, a minimum-weight $T$-join in $G$ is the (edge) disjoint union of a $v$–$w$ path and a subgraph of weight zero edges consisting of an $\emptyset$-join. Therefore, an efficient algorithm for finding a minimum-weight $T$-join has, as a special case, an efficient algorithm for finding a minimum-weight $v$–$w$ path in an undirected graph.

Using a couple of lemmata, we show how, for the problem of finding a minimum-weight $T$-join, it suffices to consider the case of nonnegative weights.

**Lemma (Symmetric difference for $T$-joins).** *Let $F'$ be a $T'$-join of $G$; let $F \subset E(G)$ and $T \subset V(G)$. Then $S$ is a $T$-join of $G$ if and only if $F \Delta F'$ is a $T \Delta T'$-join of $G$.*

*Proof.* In what follows, all arithmetic is in **GF(2)**. Recall that the statement that $F'$ (respectively, $F$, $F \Delta F'$) is a $T'$-join (respectively, $T$-join, $T \Delta T'$-join) is equivalent to the statement that $A(G)x(F') = x(T')$ [respectively, $A(G)x(F) = x(T)$, $A(G)x(F \Delta F') = x(T \Delta T')$]. Because $x(F \Delta F') = x(F) + x(F')$, and $x(T \Delta T') = x(T) + x(T')$, we have that $A(G)x(F \Delta F') = x(T \Delta T')$ is equivalent to $A(G)x(F) + A(G)x(F') = x(T) + x(T')$. The result follows. ∎

**Lemma (Shifting the objective for $T$-joins).** *Let $F'$ be a $T'$ join of $G$, and let $c$ be a weight function on $E(G)$. Define a new weight function $d$ on $E(G)$ by $d(F) := c(F \Delta F') - c(F')$, for all $F \subset E(G)$. Then $F$ is a $T$-join maximizing $d(F)$ if and only if $F \Delta F'$ is a $T \Delta T'$-join maximizing $c(F \Delta F')$.*

*Proof.* By the previous lemma, $F$ is a $T$-join of $G$ if and only if $F \Delta F'$ is a $T \Delta T'$-join of $G$. Moreover, the objective function $d(F)$ differs from $c(F \Delta F')$ by the *constant* $c(F')$. The result follows. ∎

**Theorem (Transformation to nonnegative weights for $T$-joins).** *Let $E_- := \{e \in E(G) : c(e) < 0\}$. Let $\mathcal{O}_- := \{v \in V(G) : |E_- \cap \delta_G(v)| \text{ is odd}\}$. We define a nonnegative-weight function $c^+$ on $E(G)$ by simply letting $c^+(e) := |c(e)|$, for all $e \in E(G)$. Then $F \Delta E_-$ is a minimum-weight $T \Delta \mathcal{O}_-$-join with respect to $c$ if and only if $F$ is a minimum-weight $T$-join with respect to $c^+$.*

*Proof.* We use the previous lemmata, taking $T' := E_-$, and $F' := \mathcal{O}_-$. It is easy to verify that $F'$ is a $T'$-join. Also, $d(F) := c(F \Delta F') - c(F') = c(F \setminus F') + c(F' \setminus F) - c(F') = c(F \setminus F') - c(F' \cap F) = c^+(F)$. Hence, the result follows. ∎

We return to the problem of finding a minimum-weight $T$-join of a graph $G$, for which we now assume that the weight function $c$ is nonnegative.

**Lemma (Structure of repeated edges).** *Let $P$ be a minimal $T$-join of $G$. Then $P$ partitions into the edge sets of paths* (1) *that do not share endpoints, and* (2) *whose endpoints are in $T$.*

*Proof.* The proof is by induction on the number of edges in $P$. Choose a nontrivial component $H$ of $P$. Such a component has no cycle, so it contains (at least) two vertices of degree 1. These vertices must be in $T$. There is a (unique) path in $P$ between these two vertices. Remove the edges of this path from $P$, and the result follows by induction. ■

This lemma validates the following algorithm.

**Edmonds–Johnson Minimum-Weight $T$-Join Algorithm**

0. Given graph $G$, even-cardinality subset $T$ of $V(G)$, and a nonnegative-weight function $c$ on $E(G)$.
1. For distinct $i, j \in T$, let $P_{\{i,j\}}$ be a minimum-weight $i$–$j$ path in $G$. Let $K$ be a complete graph having $V(K) := T$. Let

$$c'(\{i, j\}) := \sum_{e \in P_{\{i,j\}}} c(e), \qquad \forall \{i, j\} \in E(K).$$

2. Let $S$ be a minimum-weight perfect matching of $K$ with respect to the weight function $c'$. Then

$$P := \text{the symmetric difference of the } P_{\{i,j\}}, \text{ over } \{i, j\} \in S,$$

is a minimum-weight $T$-join of $G$.

$T$-joins also have a use in certain routing problems. These applications use the notion of "Eulerian graphs." An *Eulerian tour* of an undirected graph $G$ is a directed walk (for some orientation of the edges of $G$) that contains each element of $E(G)$ exactly once. An undirected graph is *Eulerian* if it has an Eulerian tour.

**Euler's Theorem.** *$G$ is Eulerian if and only if $G$ is connected and $E(G)$ is an $\emptyset$-join.*

*Proof.* The "if" part is obvious because an Eulerian tour must contain all of $E(G)$, and every time a walk passes through a vertex $v$, two edges meeting $v$ must be utilized.

Next, we demonstrate the "only if" part. We proceed by induction on the number of edges. Suppose that $G$ is connected and $E(G)$ is an $\emptyset$-join. It is easy to see that $E(G)$ contains a cycle $C$ of $G$. Now, the connected components $G_1, G_2, \ldots, G_k$ of $G.(E(G) \setminus C)$ also have even degree at every vertex. By the inductive hypothesis, each has an Eulerian tour. We can traverse the cycle $C$, taking "side trips" of Eulerian tours of each of the $G_i$ to construct an Eulerian tour of $G$. ■

Let $G$ be a connected graph with a nonnegative-weight function $c$ on $E(G)$. A *postperson's tour* of $G$ is a walk of $G$ that contains every edge, with repetitions allowed, and returns to its starting point. We may think of a postperson's tour of $G$ as an Eulerian tour of an Eulerian graph $\widehat{G}$ having $V(G) = V(\widehat{G})$ and $E(G) \subset E(\widehat{G})$.

**Lemma (Forest of repeated edges).** *A minimum-weight postperson's tour $\widehat{G}$ of $G$ can be chosen so that $E(\widehat{G}) \setminus E(G)$ contains no cycle. In particular, a minimum-weight postperson's tour of $G$ need not traverse any edge of $G$ more than twice.*

*Proof.* Removing any cycle from $E(\widehat{G}) \setminus E(G)$ preserves the Eulerian property (recall that $G$ is assumed to be connected). The result follows because every cycle has nonnegative weight. ■

Let $G^{(2)}$ be the weighted graph obtained from $G$ by the duplication of each edge of $G$. For specificity, let $G'$ denote the graph having $V(G') := V(G)$ and $E(G') := E(G^{(2)}) \setminus E(G)$.

As a consequence of the lemma, we can recast the problem of finding a minimum-weight postperson's tour problem as that of finding an Eulerian graph $\widehat{G}$ that has minimum edge weight, such that $\widehat{G}$ is a restriction of $G^{(2)}$ and $G$ is a restriction of $\widehat{G}$. Let $T$ be the set of vertices of $G$ having odd degree. In the language of $T$-joins, for the sought-after $\widehat{G}$, $\widehat{G} \setminus G \subset G'$ is a minimum-weight $T$-join of $G'$.

**Exercise (Postperson's tour).** Find a minimum-weight postperson's tour for the following graph:

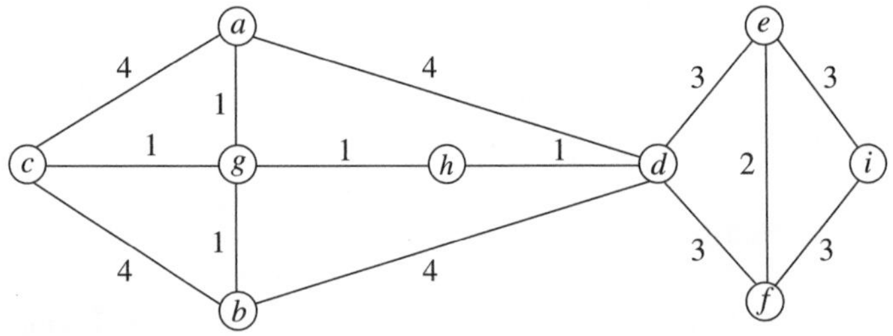


(Solve the needed minimum-weight perfect-matching problem by inspection.)

Similar ideas can be used in a heuristic for a much harder routing problem. A *Hamiltonian tour* of the undirected graph $G$ is a directed Hamiltonian tour for some orientation of the edges of $G$. Let $N$ be a finite set of distinct points in some metric space. Let $G$ be a *complete* graph with $V(G) = N$. For $e \in E(G)$, let $c(e)$ be the distance between the endpoints of $e$. *The Metric Traveling-Salesperson's Problem* is the problem of finding a minimum-weight Hamiltonian tour of $G$. For any $H \subset E(G^{(2)})$ such that $G^{(2)}.H$ is Eulerian, we easily construct a Hamiltonian tour of $G$ that has weight no greater than a given Eulerian tour of $G^{(2)}.H$, by "compressing" the tour (taking advantage of the "triangle inequality" for $c$) – that is, removing repeated interior vertices from the imputed vertex sequence of the Eulerian tour, and then taking the (unique) Hamiltonian tour with this imputed vertex sequence.

Consider the following method for determining a Hamiltonian tour of $G$ that has weight no more than 50% greater than that of a minimum-weight Hamiltonian tour.

**Christofides's Heuristic**

1. Let $S$ be a minimum-weight spanning tree of $G$, and let

$$T := \{v \in V(G) \ : \ |\delta_G(v) \cap S| \text{ is odd}\}.$$

2. Let $F$ be a minimum-weight perfect matching of $G'[T]$.
3. Define a subgraph $H$ of $G^{(2)}$, where $V(H) := V(G)$ and $E(H)$ consists of $S \cup F$. Find an Eulerian tour of $H$.
4. Compress the Eulerian tour to a Hamiltonian tour of $G$.

**Christofides's Theorem.** *Christofides's Heuristic finds a Hamiltonian tour of $G$ that has weight no more than 50% greater than that of a minimum-weight Hamiltonian tour.*

*Proof.* Because every Hamiltonian tour contains a spanning tree, the weight of $S$ is no more than the weight of a minimum-weight Hamiltonian tour. The triangle inequality implies that the weight of a minimum-weight Hamiltonian tour of $G[T]$ is no more than the weight of a minimum-weight Hamiltonian tour of $G$. Notice that $|T|$ is even. The edge set of every Hamiltonian tour of $G[T]$ is the disjoint union of two perfect matchings of $G[T]$. Therefore, the weight of $F$ is no more than 50% of the weight of a minimum-weight Hamiltonian tour of $G[T]$. Therefore, we can conclude that the weight of $E(H)$ is no more than 50% greater than that of a minimum-weight Hamiltonian tour of $V(G)$. By Euler's Theorem, $H$ has an Eulerian tour, and compressing such a tour provides a Hamiltonian tour with weight no greater than that of the Eulerian tour. ∎

**Example (Christofides's Heuristic).** Points:

Minimum-weight spanning tree of $G$:

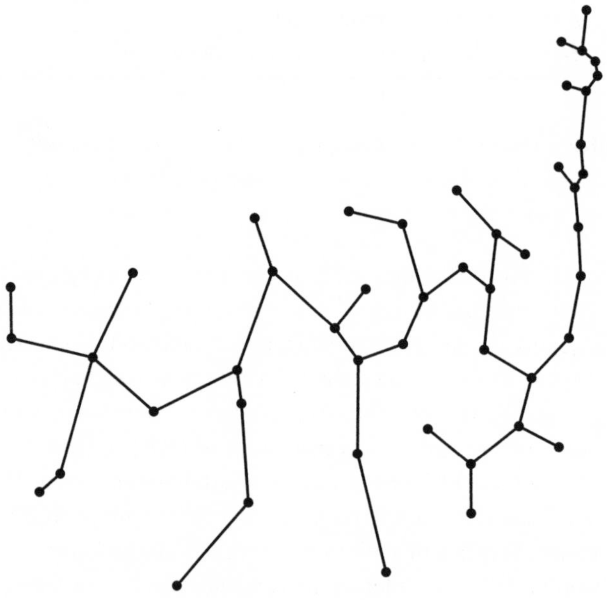

Minimum-weight perfect matching of $G'[T]$:

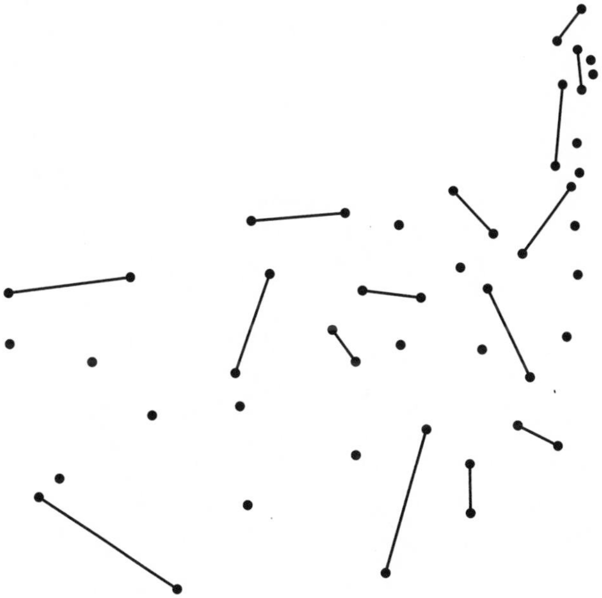

Matching and tree together (double edges rendered thicker):

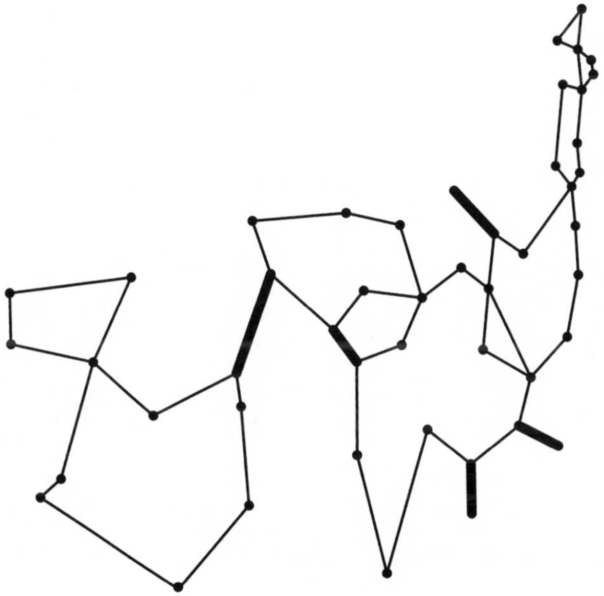

Tour generated by Christofides's Heuristic:

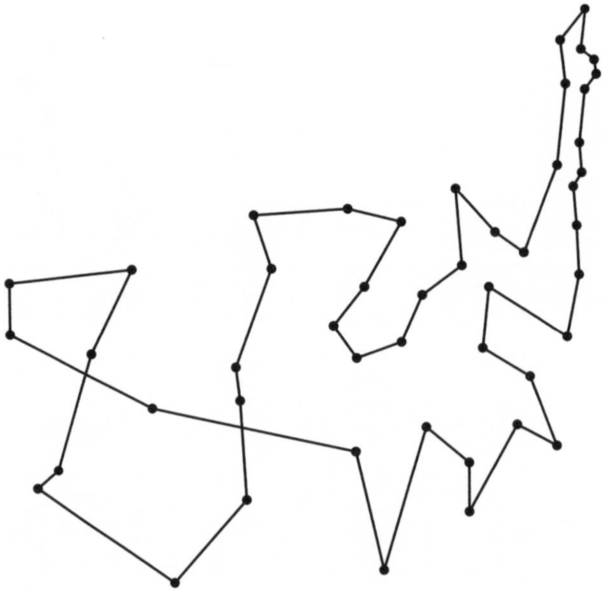

The minimum-weight Hamiltonian tour turns out to be

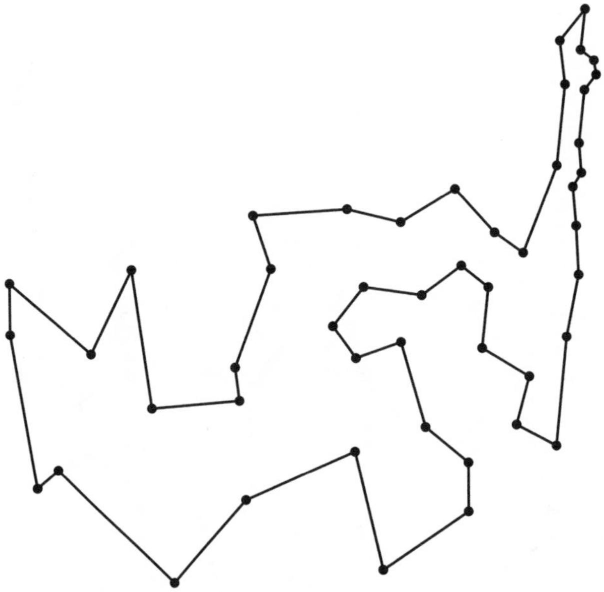

For this example, the tour generated by Christofides' heuristic weighs less than 8% more than the optimal tour – much better than the 50% guarantee. ♠

**Problem (Worst case for Christofides's Heuristic).** Show how, for every $\epsilon > 0$, there exist points in two-dimensional Euclidean space, so that Christofides's Heuristic can find a Hamiltonian tour that has weight at least $(50 - \epsilon)\%$ greater than that of a minimum-weight Hamiltonian tour. *Hint:* For $m \geq 2$, consider the $2m - 1$ points of the following "trellis." All of the triangles are identical equilateral triangles. Although lines have not been drawn between every pair of points, the distance between *every* pair of points is their Euclidean distance.

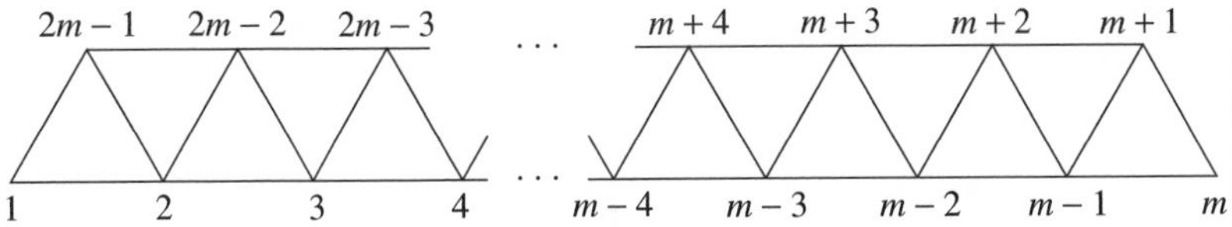


With regard to the minimum-weight Hamiltonian tour problem, there is another use for optimal-weight matchings. A set $S \subset E(G)$ is a *2-factor* of $G$ if $|S \cap \delta_G(v)| = 2$ for all $v \in V(G)$. Clearly every Hamiltonian tour is a 2-factor, so the weight of a minimum-weight 2-factor is a lower bound on the minimum-weight of a Hamiltonian tour. Next, a description is given of how to reduce the problem of finding a minimum-weight 2-factor of $G$ to the problem of finding a minimum-weight perfect matching of another graph, $G''$. In what follows, *vertex requirement* means the number of edges that must be chosen to be adjacent to a vertex. For a 2-factor, the vertex requirements are all 1; for a perfect matching, the vertex requirements are all 1. Therefore, our goal is to transform the problem on $G$ with vertex requirements of 2 to a problem on a new graph $G''$ with vertex requirements of 1. We accomplish this in two steps. The first step is to transform the problem to one, on a graph $G'$, in which all of the vertices with a vertex requirement of 2 are nonadjacent. Then we make a further transformation so that all of the vertex requirements are 1.

First, to form the problem on the graph $G'$, we take each edge $\{v, w\}$ of $G$

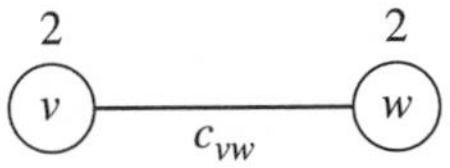

having weight $c_{vw}$ and vertex requirements 2 at $v$ and $w$, and replace it with the path

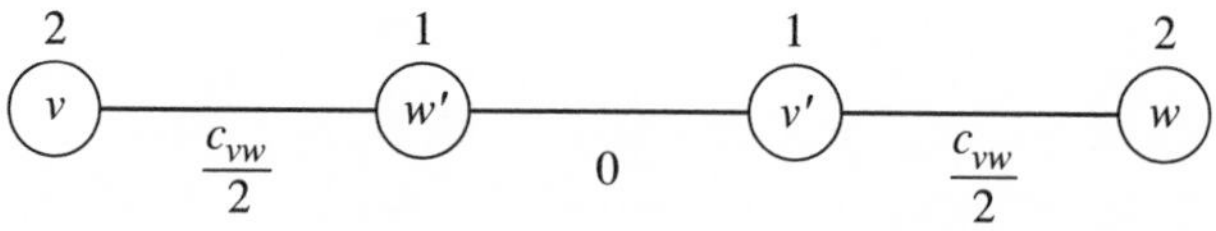


with edge weights $c_{vw'} = c_{v'w} = c_{vw}/2$ and $c_{w'v'} = 0$, and vertex requirements 2 at $v$ and $w$ and 1 at $w'$ and $v'$. In this manner, we make the following correspondence between a feasible $S$ in $G$ and a feasible $S'$ in $G'$:

$$\{v, w\} \in S \iff \{v, w'\}, \{w, v'\} \in S',$$
$$\{v, w\} \notin S \iff \{v', w'\} \in S'.$$

Next, we take each vertex $v$

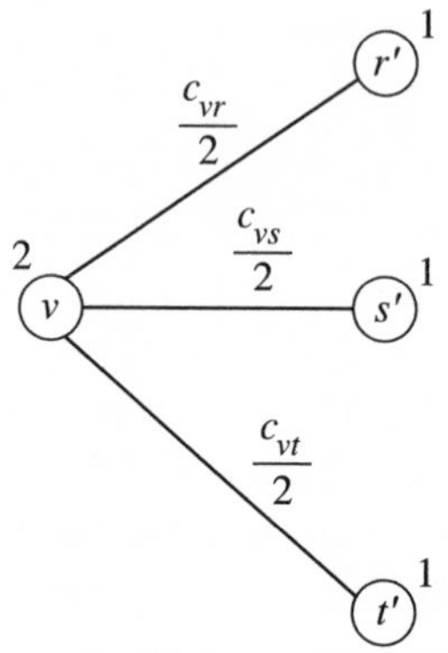


with requirement 2 and replace it with a pair of vertices, $v, \overline{v}$,

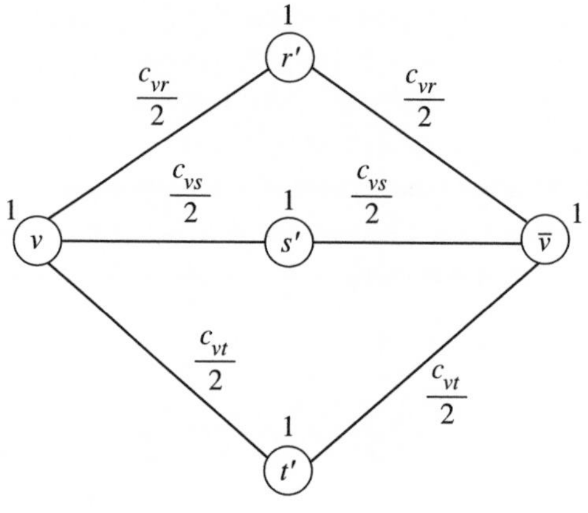

each connected to the neighbors of $v$ and with weights and vertex requirements as indicated.

## 4.6 Further Study

Another application of $T$-joins is to a minimum-weight cut problem for planar (undirected) graphs (see Section 5.3).

The definitive resource for studying matching is the monograph by Lovász and Plummer (1986). In particular, efficient algorithms for the maximum-weight matching problem can be found in that book.

Much more on the minimum-weight Hamiltonian tour problem is available in Lawler, Lenstra, Rinnooy Kan and Shmoys (1985).

Christofides's Algorithm is just a glimpse at the rich subject of approximation algorithms. The book by Vazirani (2001) is an excellent text on the subject.

# 5

## *Flows and Cuts*

An extremely useful modeling concept in the application of optimization methods to logistics problems is the idea of a flow on a digraph. A fundamental problem involving flows is the so-called maximum-flow problem. Although not a true combinatorial-optimization problem, the problem of finding a maximum flow in a digraph has a strong connection with combinatorial optimization through duality.

### 5.1 Source–Sink Flows and Cuts

Let $G$ be a digraph with no loops. We distinguish a *source* vertex $v$ and a *sink* vertex $w$ ($v \neq w$). A $v$–$w$ *flow* is a point $x \in \mathbf{R}^{E(G)}$ that satisfies the *flow-conservation equations*

$$\sum_{e \in \delta_G^+(u)} x_e - \sum_{e \in \delta_G^-(u)} x_e = 0, \quad \forall\, u \in V(G) \setminus \{v, w\}.$$

We are given an upper-bound function $c : E(G) \mapsto \mathbf{R} \cup \{+\infty\}$ and a lower-bound function $l : E(G) \mapsto \mathbf{R} \cup \{-\infty\}$ . A $v$–$w$ flow is *feasible* if it satisfies the *flow bounds*

$$l(e) \leq x_e \leq c(e), \quad \forall\, e \in E(G).$$

If we add up all flow-conservation equations, we obtain

$$-\sum_{e \in \delta_G^+(v)} x_e + \sum_{e \in \delta_G^-(v)} x_e - \sum_{e \in \delta_G^+(w)} x_e + \sum_{e \in \delta_G^-(w)} x_e = 0,$$

or, equivalently,

$$z(x) := \sum_{e \in \delta_G^+(v)} x_e - \sum_{e \in \delta_G^-(v)} x_e = \sum_{e \in \delta_G^-(w)} x_e - \sum_{e \in \delta_G^+(w)} x_e.$$

That is, the net flow out of $v$ is equal to the net flow into $w$. The *maximum $v$–$w$ flow problem* is to find a feasible $v$–$w$ flow $x$ that maximizes $z(x)$.

It is a simple observation that the maximum $v$–$w$ flow problem is a linear-programming problem. Furthermore, because the constraint matrix of the linear program is totally unimodular, it follows that, if $c : E(G) \mapsto \mathbf{Z} \cup \{+\infty\}$ and $l : E(G) \mapsto \mathbf{Z} \cup \{-\infty\}$, then there will be an integer-valued optimal solution whenever an optimal solution exists. Our goal in this section is to describe an efficient algorithm for the maximum $v$–$w$ flow problem. As a byproduct, we will have another proof of the integrality theorem previously mentioned.

It might appear that the maximum $v$–$w$ flow problem is not truly a combinatorial-optimization problem. This observation is correct, but there is a natural dual to the maximum $v$–$w$ flow problem that is a true combinatorial-optimization problem. Our efficient algorithm for the maximum $v$–$w$ flow problem will solve this dual problem as well.

A *$v$–$w$ cutset* is a set $S \subset V(G)$ such that $v \in S$ and $w \in V(G) \setminus S$. The *capacity* of $S$ is defined as

$$C(S) := \sum_{e \in \delta_G^+(S)} c(e) - \sum_{e \in \delta_G^-(S)} l(e).$$

**Lemma (Flow across a cut).** *If $x$ is a feasible $v$–$w$ flow and $S$ is a $v$–$w$ cutset, then*

$$z(x) = \sum_{e \in \delta_G^+(S)} x_e - \sum_{e \in \delta_G^-(S)} x_e.$$

*Proof.* Add up the flow-conservation equations for $u \in S - v$, and then add in the equation

$$\sum_{e \in \delta_G^+(v)} x_e - \sum_{e \in \delta_G^-(v)} x_e = \sum_{e \in \delta_G^-(w)} x_e - \sum_{e \in \delta_G^+(w)} x_e$$

to obtain

$$\sum_{e \in \delta_G^+(S)} x_e - \sum_{e \in \delta_G^-(S)} x_e = \sum_{e \in \delta_G^-(w)} x_e - \sum_{e \in \delta_G^+(w)} x_e.$$

The right-hand side is $z(x)$. ∎

**Corollary (Weak duality for flows).** *If $x$ is a feasible $v$–$w$ flow and $S$ is a $v$–$w$ cutset, then $z(x) \le C(S)$.*

*Proof.* This follows from the "Flow across a cut" Lemma, because

$$\sum_{e \in \delta_G^+(S)} x_e - \sum_{e \in \delta_G^-(S)} x_e \le C(S). \qquad \blacksquare$$

## 5.2 An Efficient Maximum-Flow Algorithm and Consequences

Let $P$ be a $v$–$w$ path. Let $P_+$ denote the forward edges of the path, and let $P_-$ denote the reverse edges of the path. A $v$–$w$ path is *augmenting* with respect to a feasible $v$–$w$ flow if

$$x_e < c(e), \quad \forall\, e \in P_+;$$
$$x_e > l(e), \quad \forall\, e \in P_-.$$

**Lemma (Augmenting paths).** *If $P$ is an augmenting $v$–$w$ path with respect to $x$, then $x$ is not a maximum $v$–$w$ flow.*

*Proof.* Let $x' \in \mathbf{R}^E$ be defined by

$$x'_e := \begin{cases} x_e + \Delta, & \text{for } e \in P_+ \\ x_e - \Delta, & \text{for } e \in P_- \\ x_e, & \text{for } e \in E(G) \setminus P \end{cases}.$$

Clearly $x'$ is a feasible flow for sufficiently small $\Delta > 0$. We find that the result follows by noting that $z(x') = z(x) + \Delta$. ∎

Ford and Fulkerson's algorithm for finding a maximum $v$–$w$ flow is motivated by the proof of the preceding lemma. At each step of the algorithm, a feasible $v$–$w$ flow $x$ is at hand, and we seek an augmenting $v$–$w$ path $P$ with respect to $x$. If we find such an augmenting path $P$, then we adjust the flow to $x'$ of the proof, taking $\Delta$ to be as large as possible subject to $x'$ satisfying the flow bounds. If an edge $e \in P$ has $x'_e = l(e)$ or $c(e)$, then $e$ is *critical* for the augmentation. If there is no maximum value for $\Delta$, then there is no maximum value for the flow. If there is no augmenting path with respect to $x$, we demonstrate that $x$ is a maximum $v$–$w$ flow by constructing a $v$–$w$ cutset $S$ with $z(x) = C(S)$.

**Maximum-Flow Algorithm**

1. Let $x$ be a feasible $v$–$w$ flow.
2. Choose an augmenting $v$–$w$ path $P$. If no such path exists, then stop.
3. If $P$ is an augmenting $v$–$w$ path, let

$$\Delta := \min\{\{c(e) - x_e \ : e \in P_+\} \cup \{x_e - l(e) \ : e \in P_-\}\}.$$

If $\Delta = +\infty$, then stop.

4. Let

$$x_e := \begin{cases} x_e + \Delta, & \text{for } e \in P_+ \\ x_e - \Delta, & \text{for } e \in P_- \\ x_e, & \text{for } e \in E(G) \setminus P \end{cases},$$

and go to Step 2.

Next, we specify Ford and Fulkerson's procedure for carrying out Step 2 of the Maximum-Flow Algorithm. In the algorithm, $L$ is the set of *labeled but unscanned vertices*. $S$ is the set of *scanned vertices* (which are labeled). $U$ is the set of *unscanned vertices*. The functions $\Delta : V(G) \mapsto \mathbf{R}_+ \cup \{+\infty\}$ and $\phi : (V(G) - v) \mapsto E(G)$ are referred to as *labeling functions*.

**Augmenting-Path Procedure**

1. $\Delta(v) := +\infty$. $L := \{v\}$. $S := \emptyset$. $U := V(G) - v$.
2. If $L = \emptyset$, then return (and report that no augmenting path exists). Otherwise, choose $u \in L$. $L := L - u$.
3. Scan vertex $u$ by repeating Step 3.i or 3.ii until no further labeling is possible. Return as soon as $w \in L$ (and report that an augmenting path exists).
   i. Choose $e \in \delta_G^+(u)$, such that $x_e < c(e)$ and $h(e) \in U$, let $\phi(h(e)) := e^+$ and $\Delta(h(e)) := \min\{\Delta(u), c(e) - x_e\}$, and let $U := U - h(e)$ and $L := L + h(e)$.
   ii. Choose $e \in \delta_G^-(u)$, such that $x_e > l(e)$ and $t(e) \in U$, let $\phi(t(e)) := e^-$ and $\Delta(t(e)) := \min\{\Delta(u), x_e - l(e)\}$, and let $U := U - t(e)$ and $L := L + t(e)$.
4. Let $S := S + u$ and go to Step 2.

The Augmenting-Path Procedure calculates enough information to obtain the forward edges $P_+$ and reverse edges $P_-$ of the augmenting path and the flow increment $\Delta$ required by the Maximum-Flow Algorithm. In particular, we have $\Delta := \Delta(w)$. We can recover $P_+$ and $P_-$ by the following backtracking procedure.

**Backtracking Procedure**

1. $u := w$. $P_+ := \emptyset$, $P_- := \emptyset$.
2. While $u \neq v$, carry out the appropriate Step 2.i or 2.ii:
   i. If $u = h(\phi(u))$, then $P_+ := P_+ + \phi(u)$ and $u := t(\phi(u))$.
   ii. If $u = t(\phi(u))$, then $P_- := P_- + \phi(u)$ and $u := h(\phi(u))$.

**Lemma (Strong duality for flows and the stopping criterion of the Maximum-Flow Algorithm).** *If the Maximum-Flow Algorithm terminates in Step 2, $x$ is a maximum $v$–$w$ flow, and $S$, determined by the Augmenting-Path Procedure, is a minimum $v$–$w$ cutset with $z(x) = C(S)$.*

*Proof.* If the Maximum-Flow Algorithm terminates in Step 2, it is because the Augmenting-Path Procedure terminated in step 2. This means that $w$ is unlabeled, so $S$ is a $v$–$w$ cutset. Moreover, $x_e = c(e)$ for all $e \in \delta_G^+(S)$, and $x_e = l(e)$ for all $e \in \delta_G^-(S)$. Therefore, $z(x) = C(S)$, and the results follows from the "Weak duality for flows" Lemma. ∎

**Example (Maximum-Flow Algorithm).** Each edge $j$ is labeled $l(j)/x_j/c(j)$. Each vertex $i$ is labeled $\Delta(i)\phi(i)$.

Initial flow value $z(x) = 4$:

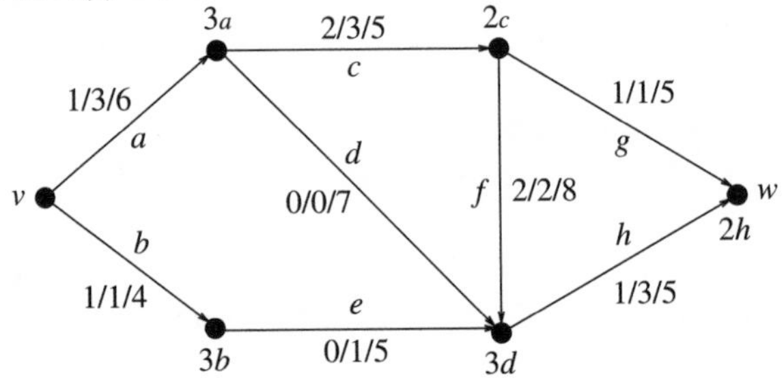


$P_+ = \{h, d, a\}$, $P_- = \emptyset$.

$z(x) = 6$:

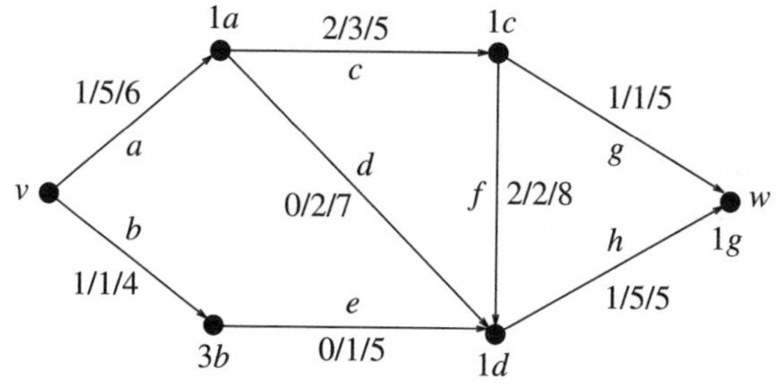


$P_+ = \{g, c, a\}$, $P_- = \emptyset$.

---

$z(x) = 7$:

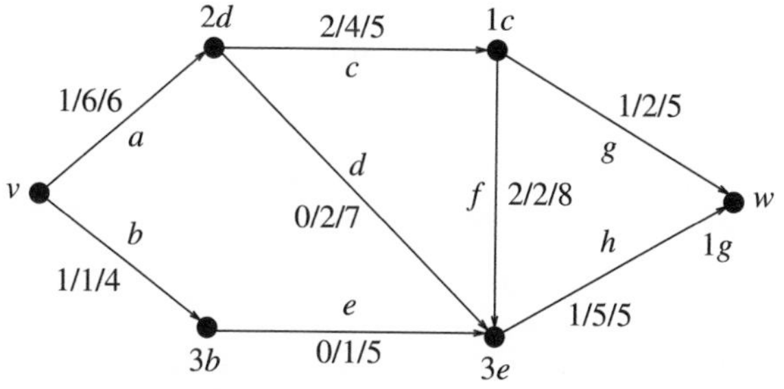


$P_+ = \{g, c, e, b\}$, $P_- = \{d\}$.

---

$z(x) = C(S) = 8$:

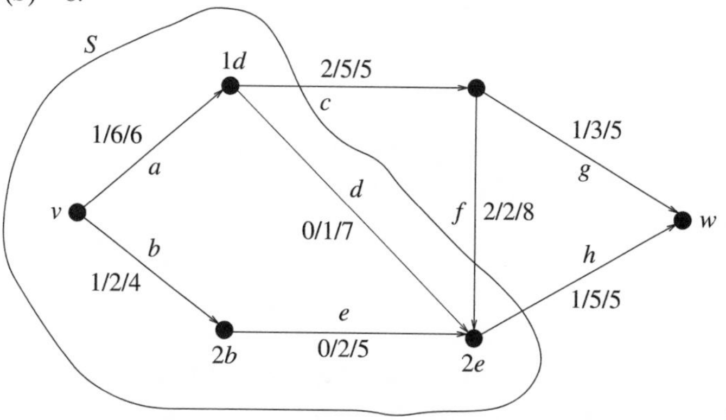


♠

**Edmonds–Karp Theorem.** *If vertices are chosen from L in Step 2 of the Augmenting-Path Procedure in the order that they were appended to L in Steps 3.i and 3.ii, then the Maximum-Flow Algorithm terminates with no more than* $|V(G)| \cdot |E(G)|/2$ *repetitions of Step 2.*

If vertices are scanned on a first-labeled/first-scanned basis, then each flow augmentation uses an augmenting path with a minimum number of edges.

**Exercise (Edmonds-Karp labeling).** Let $M$ be a large positive integer. Each edge $j$ is labeled $l(j)/x_j/c(j)$.

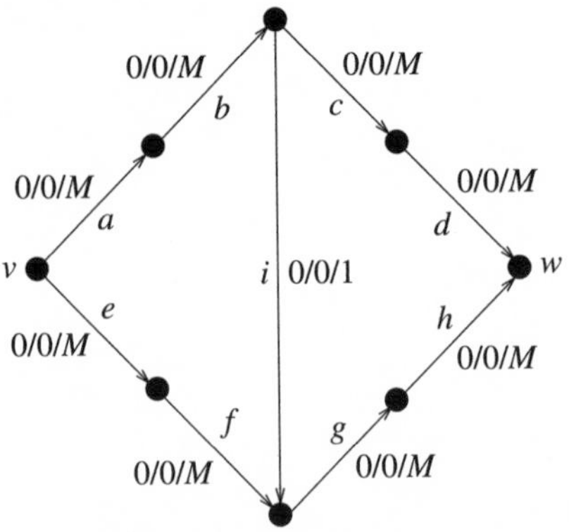


Show that, when the Edmonds–Karp labeling is used, the optimal flow is found in just two augmentations, whereas, without any special precautions being taken, the Maximum-Flow Algorithm may use $2M$ augmentations. Also, if $M = +\infty$, then the algorithm can iterate infinitely.

For all $i \in V(G)$ and nonnegative integers $k$, let

$$\sigma_i^k := \text{the minimum number of edges in a } v\text{–}i \text{ augmenting path after } k \text{ flow augmentations.}$$

and let

$$\tau_i^k := \text{the minimum number of edges in an } i\text{–}w \text{ augmenting path after } k \text{ flow augmentations.}$$

**Lemma (Monotonicity of labels in the Maximum-Flow Algorithm).** *If each flow augmentation uses an augmenting path with a minimum number of edges, then $\sigma_i^{k+1} \geq \sigma_i^k$ and $\tau_i^{k+1} \geq \tau_i^k$ for all $i \in V(G)$ and nonnegative integers $k$. In particular, the numbers of edges in the augmenting paths chosen by the algorithm never decrease.*

*Proof (Lemma).* Suppose that $\sigma_i^{k+1} < \sigma_i^k$ for some $i \in V(G)$ and some $k$. For any such $k$, we can choose $i$ so that $\sigma_i^{k+1}$ is as small as possible (among all $i$ with $\sigma_i^{k+1} < \sigma_i^k$). Obviously, $\sigma_i^{k+1} \geq 1$ (because $\sigma_v^{k+1} = \sigma_v^k = 0$).

There is some final edge $e$ on a shortest $v$–$i$ augmenting path after $k+1$ augmentations. Let us suppose that $e$ is a forward edge in the path (the case in which it is a reverse edge is handled similarly). Therefore, $x_e < c(e)$. Clearly, $\sigma_i^{k+1} = \sigma_{t(e)}^{k+1} + 1$.

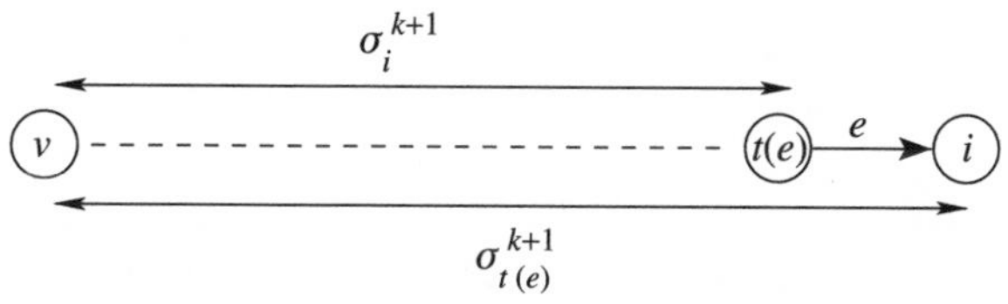


By the choice of $i$, we have $\sigma_{t(e)}^{k+1} \geq \sigma_{t(e)}^k$. Hence, $\sigma_i^{k+1} \geq \sigma_{t(e)}^k + 1$.

Now, suppose that $x_e < c(e)$ after $k$ flow augmentations. Then, we would have $\sigma_i^k \leq \sigma_{t(e)}^k + 1$. However, we have shown that $\sigma_i^{k+1} \geq \sigma_{t(e)}^k + 1$; therefore, $\sigma_i^k \leq \sigma_i^{k+1}$, which is contrary to our assumption.

Therefore, we must have that $x_e = c(e)$ after $k$ flow augmentations. However, this implies that $e$ was a reverse edge in the $k+1$st $v$–$w$ augmenting path. Because that path had the minimum number of edges, we have $\sigma_{t(e)}^k = \sigma_i^k + 1$. Now, we have already established that $\sigma_i^{k+1} \geq \sigma_{t(e)}^k + 1$. Therefore, we have that $\sigma_i^{k+1} \geq \sigma_i^k + 2$, contradicting $\sigma_i^{k+1} < \sigma_i^k$.

The proof that $\tau_i^{k+1} \geq \tau_i^k$ is very similar. ■

*Proof (Edmonds–Karp Theorem).* Suppose that edge $e$ is critical for the $k+1$st augmentation. The number of edges in the associated augmenting path is $\sigma_{t(e)}^k + \tau_{t(e)}^k = \sigma_{h(e)}^k + \tau_{h(e)}^k$, because the path has the fewest number of edges among $v$–$w$ augmenting paths. The next time that edge $e$ appears in an augmentation, say the $l+1$st augmentation, it appears with the opposite orientation. Let us suppose that $e$ is a forward edge in the $k+1$st augmentation (the case in which it is a reverse edge is handled similarly). Then, $\sigma_{h(e)}^k = \sigma_{t(e)}^k + 1$ and $\sigma_{t(e)}^l = \sigma_{h(e)}^l + 1$. Using the lemma, we have that $\sigma_{t(e)}^l - 1 = \sigma_{h(e)}^l \geq \sigma_{h(e)}^k = \sigma_{t(e)}^k + 1$; therefore, $\sigma_{t(e)}^l \geq \sigma_{t(e)}^k + 2$. The lemma also gives us $\tau_{t(e)}^l \geq \tau_{t(e)}^k$, so we can conclude that $\sigma_{t(e)}^l + \tau_{t(e)}^l \geq \sigma_{t(e)}^k + \tau_{t(e)}^k + 2$.

That is, every time an edge is critical, the length of the augmenting path that uses it is at least two edges longer than the previous time that it was critical. Now, every augmenting path has at most $|V(G)| - 1$ edges, and therefore, an edge can not be critical more than $|V(G)|/2$ times. The result follows because $G$ has $|E(G)|$ candidates for critical edges. ∎

**Corollary (Max-Flow/Min-Cut Theorem).** *The maximum value of a $v$–$w$ flow is equal to the minimum capacity of a $v$–$w$ cutset.*

**Problem (Linear-programming proof of the Max-Flow/Min-Cut Theorem).** Using linear-programming duality and total unimodularity, give another proof of the Max-Flow/Min-Cut Theorem.

**Problem (Finding a feasible $v$–$w$ flow).** Given a maximum $v$–$w$ flow problem *MFP*, show how to formulate another maximum source-sink flow problem *MFP′* so that

a. *MFP′* has a readily available feasible source-sink flow, and
b. any optimal solution of *MFP′* either reveals that *MFP* has no feasible $v$–$w$ flow or calculates a feasible $v$–$w$ flow for *MFP*.

**Problem (König's Theorem).** Let $G'$ be a (undirected) *bipartite* graph with vertex bipartition $(V_1, V_2)$. Associated with $G'$ is a directed graph $G$, with $V(G) := V(G') \cup \{v, w\}$. The edges of $G$ are the edges of $G'$, all directed from the $V_1$ side to the $V_2$ side, together with edges of the form $(v, i)$ for all $i \in V_1$, and $(i, w)$ for all $i \in V_2$. We consider flows in $G$. All edges $e$ have flow lower bounds $l(e) = 0$. All edges $e$ between $V_1$ and $V_2$ have flow upper bounds of $c(e) = +\infty$. All other edges $e$ have flow upper bounds of $c(e) = 1$.

a. Explain how finding a maximum $v$–$w$ flow in $G$ solves the problem of finding a maximum-cardinality matching of $G'$.
b. Use the Max-Flow/Min-Cut Theorem to prove König's Theorem (see p. 44) that the maximum cardinality of a matching of the bipartite graph $G'$ is equal to the minimum cardinality of a vertex cover of $G'$.

**Problem (Vertex packing in bipartite graphs).** Let $G$ be a digraph having no loops with source $v$ and sink $w$. We consider $v$–$w$ flows $x$ that respect

$l(e) \le x_e \le c(e)$. The *anticapacity* of a $v$–$w$ cutset $S$ is defined as

$$L(S) := \sum_{e \in \delta_G^+(S)} l(e) - \sum_{e \in \delta_G^-(S)} c(e) = -C(V(G) \setminus S).$$

a. Prove that the *minimum* value of a $v$–$w$ flow is equal to the *maximum* anticapacity of a $v$–$w$ cutset.

b. Let $G'$ be a (undirected) *bipartite* graph with vertex bipartition $(V_1, V_2)$. A set $X \subset E(G')$ is an *edge cover* of $G'$ if every element of $V(G')$ is met by some element of $X$. Associated with $G'$ is a digraph $G$, with $V(G) := V(G') \cup \{v, w\}$. The edges of $G$ are the edges of $G'$, all directed from the $V_1$ side to the $V_2$ side, together with edges of the form $(v, i)$ for all $i \in V_1$, and $(i, w)$ for all $i \in V_2$. We consider flows in $G$. All edges $e$ have capacity $c(e) = +\infty$. All edges $e$ between $V_1$ and $V_2$ have flow lower bounds of $l(e) = 0$. All other edges $e$ have flow lower bounds of $l(e) = 1$. Explain how finding a *minimum* $v$–$w$ flow in $G$ solves the problem of finding a minimum-cardinality edge cover of $G'$.

c. Assume that $G'$ has no isolated vertices. Prove that the minimum cardinality of an edge cover of $G'$ is equal to the maximum cardinality of a vertex packing of $G'$.

## 5.3 Undirected Cuts

For an *un*directed graph $H$ with $d : E(H) \mapsto \mathbf{R}_+$, it is a simple matter to find a minimum-weight $v$–$w$ cut, that is, to solve

$$\min_{S \subset V(H)} \{f(S) \; : \; v \in S, \; w \in V(H) \setminus S\},$$

where $f(S) := \sum_{e \in \delta_H(S)} d(e)$. We simply make a digraph $G$ having $V(G) := V(H)$. Each edge $\{i, j\}$ of $H$ gives rise to an oppositely directed pair of edges $e'$ and $e''$ in $G$ [i.e., with $t(e') := h(e'') := i$ and $h(e') := t(e'') := j$]. Then we define an upper-bound function $c$ on $E(G)$ by letting $c(e') := c(e'') := d(e)$ and a lower-bound function $l$ on $E(G)$ by letting $c(e') := c(e'') := 0$. Then the capacity $C(S)$ of each $v$–$w$ cut $S$ in graph $G$ is precisely equal to $f(S)$. Therefore, we can find a minimum $v$–$w$ cut in the undirected graph $H$ by using an efficient algorithm to find a minimum $v$–$w$ cut in the digraph $G$.

On the other hand, if the weight function $d$ is not nonnegative, then the preceding transformation does not work. In that case, the problem is considerably harder. In Chapter 6 we will study an integer-programming approach to the general problem (see p. 175).

For now, we describe an efficient method when the graph $H$ is planar and the source and sink are not fixed. That is, we wish to solve

$$\min_{S \subset V(H)} \{f(S)\}.$$

We rely on planar duality (see p. 65) and $T$-join methodology (see Section 4.4). A *cut* in $G$ is just a set $\delta_G(S)$ for some $S \subset V(G)$.

**Theorem ($T$-joins in the planar dual and cuts).** *Every cut in $G$ is an $\emptyset$-join in $G^*$, and conversely.*

*Proof.* Let $S$ be a subset of $V(G)$. Consider the cut $\delta_G(S)$. The graph $G'$ having $V(G') := V(G)$ and $E(G') := \delta_G(S)$ is bipartite. Therefore, $G'$ has no odd cycles. Hence, the dual of $G'$ has all even degrees. In addition, no edge of $E(G) \setminus E(G')$ connects two nonadjacent elements of $V(G)$. Therefore, the set of dual edges of $E(G')$ is a subset of the edges of $G^*$. Hence, $E(G')$ is an $\emptyset$-join of $G^*$.

Conversely, let $F$ be an $\emptyset$-join of $G^*$. We can view $F$ as the edge-disjoint union of simple cycles. Thinking of the planar embedding of $G^*$, we can consider each cycle as a simple closed curve, with curves intersecting on vertices only. In the following example, we view the embedding of $F$ as five such curves. Each vertex of $G$ is inside or outside each such curve – regions enclosed by an odd number of these curves are shaded in the example:

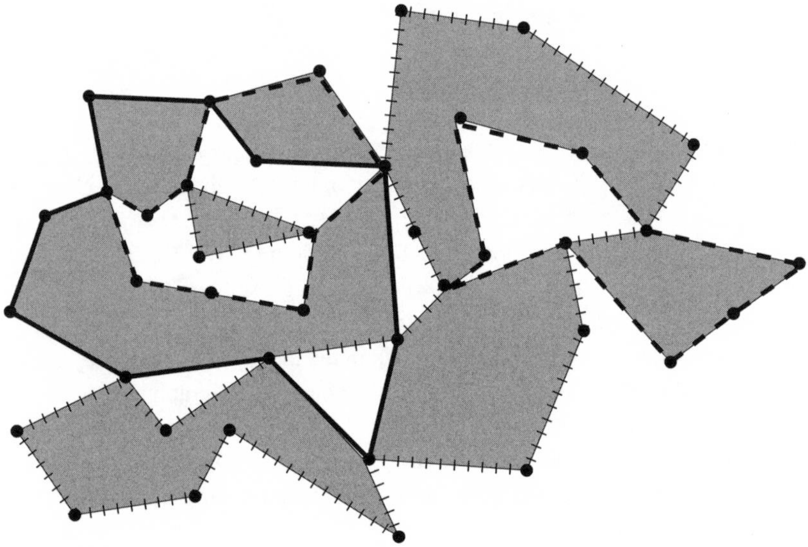

The following five "thumbnail" drawings are meant to clarify which edges of $F \subset E(G^*)$ are in each curve.

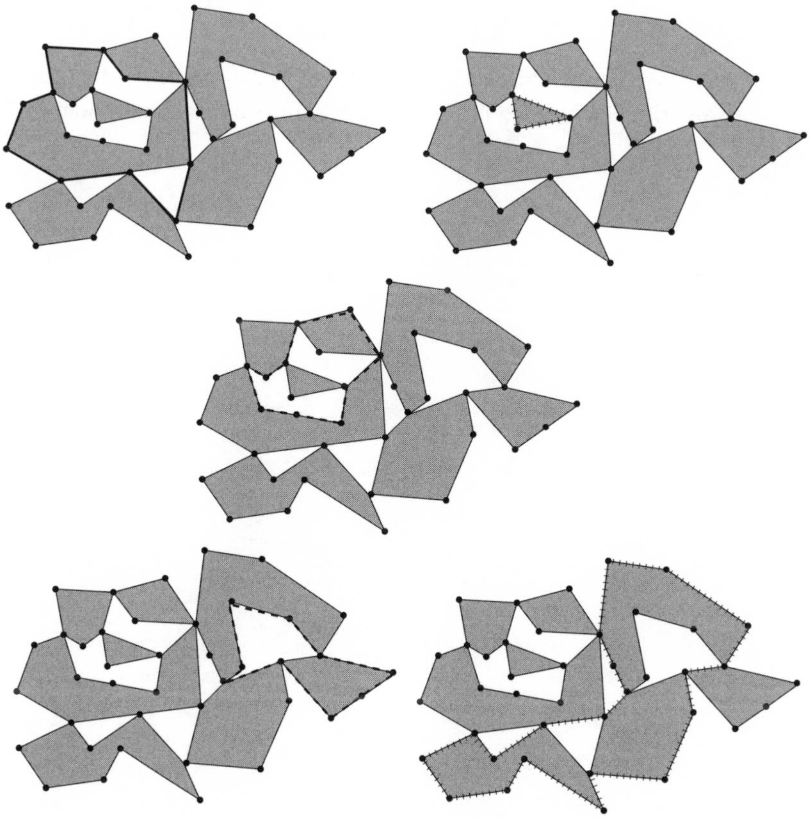

Let $S$ be the set of vertices of $G$ (not drawn) that are inside an odd number of these curves. It is easy to see that the elements of $\delta_G(S)$ are precisely the (dual) edges of $F$. ■

Therefore, we can find a minimum-weight cut of $G$ by simply taking a minimum-weight $\emptyset$-join of $G^*$ – using the transformation technique of the proof of the "Transformation to nonnegative weights for $T$-joins" Theorem (see p. 128) and the Edmonds–Johnson Minimum-Weight $T$-Join Algorithm (see p. 129). If we really want a minimum-weight $v$–$w$ cut for a particular pair of distinct vertices $v, w \in V(G)$, then we can just append an edge $\{v, w\}$ to $H$ with a large negative weight; then, as long as appending this edge leaves the graph planar, a minimum-weight cut of this new graph will be a minimum $v$–$w$ cut of $H$.

## 5.4 Further Study

The work of Ahuja, Magnanti and Orlin (1993) is an excellent reference on network algorithms. In particular, Chapters 6–8 focus on maximum-flow algorithms.

# 6

## Cutting Planes

Many combinatorial-optimization problems have natural formulations as integer linear programs. The feasible region of such a formulation may have extreme points with fractional components, in which case the optimal solution may not be obtained by solving the linear-programming relaxation. However, if we had a concise description of the convex hull of the feasible points, by means of linear inequalities, we could solve such an integer linear program by solving a linear program. Thus we have a strong interest in finding such descriptions.

If we do not have such a concise description, all is not lost; as long as we can generate some of the needed inequalities in an efficient manner, we may be able to solve the integer linear program as a sequence of linear programs. A *cutting plane* is a linear inequality that is generated as needed in the course of solving an integer linear program as a sequence of linear programs. In this chapter, we study formulation and algorithmic issues involving cutting planes.

### 6.1 Generic Cutting-Plane Method

Consider the integer linear program *IP*:

$$\max \sum_{j=1}^{k} c_j x_j$$

subject to:

$$(i) \qquad \sum_{j=1}^{k} a_{ij} x_j \le b_i\,, \quad \text{for } i = 1, 2, \ldots, m;$$

$$(ii) \qquad x_j \ge 0\,, \quad \text{for } j = 1, 2, \ldots, k;$$

$$(iii) \qquad x_j \in \mathbf{Z}\,, \quad \text{for } j = 1, 2, \ldots, k.$$

We consider approaches to solving *IP* that rely on solving a sequence of linear programs that provide successively better approximations to *IP*.

**Generic Cutting-Plane Method**

0. Initially, let *LP* be the linear-programming relaxation of *IP*.
1. Let $x^*$ be an optimal extreme-point solution of *LP*.
2. If $x^*$ is all integer, then stop because $x^*$ is optimal to *IP*.
3. If $x^*$ is not all integer, then find an inequality that it satisfied by all feasible solutions of *IP*, but is violated by $x^*$, append the inequality to *LP*, and go to Step 1.

As we keep appending inequalities to our linear-programming relaxation, the sequence of optimal values to the successive linear programs is a nonincreasing sequence of upper bounds on the optimal value of *IP*. The difficulty in applying the Generic Cutting-Plane Method lies in the problem of finding the inequalities of Step 3.

## 6.2 Chvátal–Gomory Cutting Planes

Choose $u \in \mathbf{R}^m_+$. For all $x$ that satisfy the inequalities $(i)$, we have

$$(i') \qquad \sum_{i=1}^{m} u_i \sum_{j=1}^{k} a_{ij} x_j \le \sum_{i=1}^{m} u_i b_i,$$

or, equivalently,

$$\sum_{j=1}^{k} \left\lfloor \sum_{i=1}^{m} u_i a_{ij} \right\rfloor x_j + \sum_{j=1}^{k} \left( \sum_{i=1}^{m} u_i a_{ij} - \left\lfloor \sum_{i=1}^{m} u_i a_{ij} \right\rfloor \right) x_j \le \sum_{i=1}^{m} u_i b_i.$$

Therefore, we have that all solutions of $(i)$ and $(ii)$ satisfy

$$(ii') \qquad \sum_{j=1}^{k} \left\lfloor \sum_{i=1}^{m} u_i a_{ij} \right\rfloor x_j \le \sum_{i=1}^{m} u_i b_i.$$

It is important to note that we have not used $(iii)$ yet. Finally, $(iii)$ implies that all solutions of $(ii')$ and $(iii)$ satisfy the *Chvátal–Gomory cutting plane*:

$$(iii') \qquad \sum_{j=1}^{k} \left\lfloor \sum_{i=1}^{m} u_i a_{ij} \right\rfloor x_j \le \left\lfloor \sum_{i=1}^{m} u_i b_i \right\rfloor .$$

To appreciate the limitations of Chvátal–Gomory cutting planes, it is important to note that $(iii')$ must be satisfied by all solutions of $(ii')$ and $(iii)$, even those that do not satisfy $(i)$ and $(ii)$.

**Example (Chvátal–Gomory cutting planes).** Consider the program

$$\begin{aligned} &\max\ 2x_1 + x_2 \\ &\text{subject to:} \\ &\qquad 7x_1 + x_2 \le 28; \\ &\qquad -x_1 + 3x_2 \le 7; \\ &\qquad -8x_1 - 9x_2 \le -32; \\ &\qquad x_1, x_2 \ge 0; \\ &\qquad x_1, x_2 \in \mathbf{Z}. \end{aligned}$$

The choice of $u_1 = 0$, $u_2 = 1/3$, $u_3 = 1/3$ yields the cutting plane $-3x_1 -2x_2 \le -9$. The choice of $u_1 = 1/21$, $u_2 = 7/22$, $u_3 = 0$ yields the cutting plane $x_2 \le 3$.

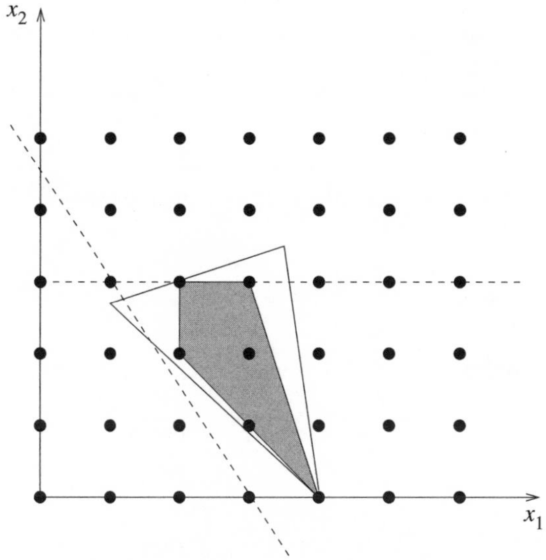

Notice how a Chvátal–Gomory cutting plane can "hang up" on integer points whether they are feasible or not. ♠

**Exercise (Chvátal–Gomory cutting planes).** Find a choice of $u_1, u_2, u_3 \geq 0$ to yield the cutting plane $-x_1 - x_2 \leq -4$ for the Chvátal–Gomory cutting-planes Example. Is there a choice of $u_1, u_2, u_3 \geq 0$ that will yield the valid inequality $-x_1 \leq -2$ as a cutting plane? *Hint:* Devise an inequality system with variables $u_1, u_2, u_3$.

**Problem (Chvátal–Gomory cutting planes).** Consider the program

$$
\begin{aligned}
\max \quad & x_2 \\
\text{subject to:} \quad & \\
2k \cdot x_1 + x_2 &\leq 2k \\
-2k \cdot x_1 + x_2 &\leq 0 \\
x_1, x_2 &\geq 0 \\
x_1, x_2 &\in \mathbf{Z},
\end{aligned}
$$

where $k$ is a positive integer. Observe that the optimal objective value of the program is 0, whereas the optimal value of the linear-programming relaxation is $k$. Graph the feasible region. Convince yourself that it would require at least $k$*-successive* Chvátal–Gomory cutting planes to reduce the optimal objective value of the linear program to 0.

**Problem (Matching and Chvátal–Gomory cutting planes).** Demonstrate how the inequalities

$$
\sum_{e \in E(G[W])} x_e \leq \frac{|W| - 1}{2}, \quad \forall\, W \subset V(G) \text{ with } |W| \geq 3 \text{ odd.}
$$

are Chvátal–Gomory cutting planes with respect to the system

$$
\begin{aligned}
\sum_{e \in \delta_G(v)} x_e &\leq 1, \quad \forall\, v \in V(G); \\
x_e &\geq 0, \quad \forall\, e \in E(G); \\
x_e &\in \mathbf{Z}, \quad \forall\, e \in E(G).
\end{aligned}
$$

**Problem (Vertex packing and Chvátal–Gomory cutting planes).** Let $G$ be a simple graph and let $P(G)$ be the convex hull of the characteristic vectors of vertex packings of $G$.

a. Demonstrate how the *clique inequalities*

$$\sum_{v \in W} x_v \leq 1, \quad \forall\, W \subset V(G) \text{ such that } G[W] \text{ is complete,}$$

arise by repeatedly applying the Chvátal–Gomory process, starting with the formulation

$$\begin{aligned} \sum_{v \in e} x_v &\leq 1, \quad \forall\, e \in E(G); \\ x_v &\geq 0, \quad \forall\, v \in V(G); \\ x_v &\in \mathbf{Z}, \quad \forall\, v \in V(G). \end{aligned}$$

b. Show that if $G[W]$ is a *maximal* complete subgraph of $G$ (i.e., $G[W]$ is complete, and there is no vertex $w \in V(G) \setminus W$ such that $G[W + w]$ is complete), then the associated clique inequality describes a facet of $P(G)$.

**Problem [Uncapacitated facility location, continued (see p. 6)].** Demonstrate how the inequalities $(**)$ are Chvátal–Gomory cutting planes with respect to the original formulation which uses $(*)$.

**Problem (Mixed-integer cuts).** Let

$$\mathcal{P} := \left\{ \begin{pmatrix} x \\ y \end{pmatrix} : \begin{aligned} & x_j \geq 0,\ j = 1, 2, \ldots, k; \\ & x_j \in \mathbf{Z},\ j = 1, 2, \ldots, k; \\ & y_j \geq 0,\ j = 1, 2, \ldots, p; \\ & \sum_{j=1}^{k} a_j x_j + \sum_{j=1}^{p} \alpha_j y_j \leq b \end{aligned} \right\}.$$

Let $\Phi := \{j \ :\ \alpha_j < 0,\ 1 \leq j \leq p\}$, and let $f := b - \lfloor b \rfloor$. Prove that the following inequality is valid for $\mathcal{P}$:

$$\sum_{j=1}^{k} \lfloor a_j \rfloor x_j + \frac{1}{1-f} \sum_{j \in \Phi} \alpha_j y_j \leq \lfloor b \rfloor .$$

*Hint:* Consider the two cases (1) $\sum_{j\in\Phi} \alpha_j y_j > f - 1$ and (2) $\sum_{j\in\Phi} \alpha_j y_j \leq f - 1$.

## 6.3 Gomory Cutting Planes

Gomory cutting planes are particular manifestations of Chvátal–Gomory cutting planes. They are "general-purpose" cutting planes in the sense that they do not rely any special structure of the integer linear program. Gomory cutting planes arise from linear-programming basic-feasible solutions. For this, we assume that the $a_{ij}$, $b_i$, and $c_j$ are all integers. Let $x_0$ be defined by

$$x_0 - \sum_{j=1}^{k} c_j x_j = 0 .$$

For $i = 1, 2, \ldots, m$, define nonnegative slack variables $x_{k+i}$ by

$$x_{k+i} + \sum_{j=1}^{k} a_{ij} x_j = b_i .$$

Note that the slack variables and $x_0$ can be forced to be integers, because the $a_{ij}$, $b_i$, and $c_j$ are all assumed to be integers (this is important!).

For $1 \leq i, j \leq m$, let

$$a_{i,k+j} := \begin{cases} 1 & \text{if } i = j \\ 0 & \text{if } i \neq j \end{cases},$$

and for $1 \leq i \leq m$ let

$$a_{i,0} := 0.$$

Finally, for $0 \leq j \leq n$, let

$$a_{0,j} := \begin{cases} 1 & \text{if } j = 0 \\ -c_j & \text{if } 1 \leq j \leq k \\ 0 & \text{if } k+1 \leq j \leq k+m \end{cases},$$

and let $b_0 := 0$.

Now, let $n := k + m$. We can represent a basic solution of the linear-programming relaxation of *IP* by manipulating the system of equations

$$\sum_{j=0}^{n} a_{ij} x_j = b_i \ , \quad \text{for } i = 0, 1, 2, \ldots, m.$$

A basic solution arises if the indices $(0, 1, 2, \ldots, n)$ are partitioned into *nonbasic indices* $\eta = (\eta_1, \eta_2, \ldots, \eta_{n-m})$ and *basic indices* $\beta = (\beta_0 := 0, \beta_1, \beta_2, \ldots, \beta_m)$ so that

$$(E_i) \qquad \sum_{j=0}^{m} a_{i\beta_j} x_{\beta_j} + \sum_{j=1}^{n-m} a_{i\eta_j} x_{\eta_j} = b_i \ , \quad \text{for } i = 0, 1, 2, \ldots, m,$$

has a unique nonnegative solution $x^*$ with $x^*_{\eta_1} = x^*_{\eta_2} = \cdots = x^*_{\eta_{n-m}} = 0$ . We can also solve the equations for the basic variables in terms of the nonbasic variables:

$$(\overline{E}_{\beta_i}) \qquad x_{\beta_i} + \sum_{j=1}^{n-m} \overline{a}_{\beta_i\eta_j} x_{\eta_j} = x^*_{\beta_i} \ , \quad \text{for } i = 0, 1, 2, \ldots, m.$$

This equation is what we ordinarily see as row $i$ of a simplex table. Using the nonnegativity and integrality of the $x_j$, we obtain the *Gomory cutting planes*

$$(G_{\beta_i}) \qquad x_{\beta_i} + \sum_{j=1}^{n-m} \lfloor \overline{a}_{\beta_i\eta_j} \rfloor x_{\eta_j} \leq \lfloor x^*_{\beta_i} \rfloor \ , \quad \text{for } i = 0, 1, 2, \ldots, m.$$

Note that the cutting plane $G_{\beta_i}$ is violated by the basic solution at hand whenever $x^*_{\beta_i} \notin \mathbf{Z}$. Furthermore, the Gomory cutting plane has integer coefficients. This last fact is important so that we can repeat the process. Finally, we note that, by subtracting $\overline{E}_{\beta_i}$ from $G_{\beta_i}$, we obtain the equivalent inequality

$$(\overline{G}_{\beta_i}) \quad \sum_{j=1}^{n-m} \left( \lfloor \overline{a}_{\beta_i\eta_j} \rfloor - \overline{a}_{\beta_i\eta_j} \right) x_{\eta_j} \leq \lfloor x^*_{\beta_i} \rfloor - x^*_{\beta_i} \ , \quad \text{for } i = 0, 1, 2, \ldots, m.$$

**Example (Gomory cutting planes).** Continuing with the Chvátal–Gomory cutting-plane example, we introduce slack variables $x_3, x_4, x_5$. The optimal linear-programming basis consists of the basic indices (0,1,2,5), as can be seen from the optimal simplex table:

$$\begin{array}{llllllll}
x_0 & & & +\frac{7}{22}x_3 & +\frac{5}{22}x_4 & & = & \frac{21}{2} \\
& x_1 & & +\frac{3}{22}x_3 & -\frac{1}{22}x_4 & & = & \frac{7}{2} \\
& & x_2 & +\frac{1}{22}x_3 & +\frac{7}{22}x_4 & & = & \frac{7}{2} \\
& & & +\frac{3}{2}x_3 & +\frac{5}{2}x_4 & +x_5 & = & \frac{55}{2}
\end{array}$$

We obtain the Gomory cutting planes

(0) $$x_0 \leq 10;$$

(1) $$x_1 - x_4 \leq 3;$$

(2) $$x_2 \leq 3;$$

(3) $$x_3 + 2x_4 + x_5 \leq 27,$$

which, in the space of the original variables, are

(0′) $$2x_1 + x_2 \leq 10;$$

(1′) $$3x_2 \leq 10;$$

(2′) $$x_2 \leq 3;$$

(3′) $$3x_1 + 2x_2 \leq 17.$$

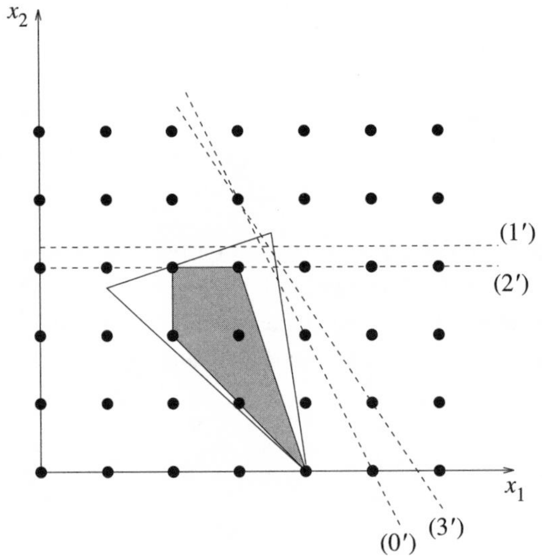


♠

> **Exercise (Gomory cutting planes).** In the example, on which integer points does inequality (1′) “hang up”?

In the space of the nonbasic variables the Gomory cutting planes take the form

$$
\begin{array}{lrrcr}
(\bar{0}) & -\frac{7}{22}x_3 & -\frac{5}{22}x_4 & \leq & -\frac{1}{2}; \\
(\bar{1}) & -\frac{3}{22}x_3 & -\frac{21}{22}x_4 & \leq & -\frac{1}{2}; \\
(\bar{2}) & -\frac{1}{22}x_3 & -\frac{7}{22}x_4 & \leq & -\frac{1}{2}; \\
(\bar{3}) & -\frac{1}{2}x_3 & -\frac{1}{2}x_4 & \leq & -\frac{1}{2}.
\end{array}
$$

**Problem (Gomory cuts are Chvátal–Gomory cuts).** Let $\lambda_{j1}, \lambda_{j2}, \ldots, \lambda_{jm}$ be real numbers such that

$$(\overline{E}_j) = \sum_{i=1}^{m} \lambda_{ji} \cdot (E_i).$$

Show how the Gomory cutting plane $G_j$ $(1 \leq j \leq m)$ is equivalent to the special case of the Chvátal–Gomory cutting plane for the choice of $u_i := \lambda_{ji} - \lfloor \lambda_{ji} \rfloor$, for $i = 1, 2, \ldots, m$. (You will need to use the observation that for $\lambda \in \mathbf{R}$, $a \in \mathbf{Z}$, $\lfloor (\lambda - \lfloor \lambda \rfloor) a \rfloor = \lfloor \lambda a \rfloor - \lfloor \lambda \rfloor a$).

Gomory's method for applying the Generic Cutting-Plane Method to solve *IP* is to select a "source" row of the simplex table having a basic variable that has a noninteger value, and generate a Gomory cutting plane from that row.

**Example [Gomory cutting planes, continued (see p. 157)].** Next, we finish solving the example by using Gomory's method. We append the Gomory cutting plane to the bottom of our simplex table after writing it in terms of the nonbasic variables and introducing a (basic) slack variable. We then solve the resulting linear program with the dual simplex method. Each "←" indicates the source row for each cutting plane. Pivot elements for the dual simplex method are indicated by "[· · ·]".

| $10\frac{1}{2}$ | $3\frac{1}{2}$ | $3\frac{1}{2}$ | 0 | 0 | $27\frac{1}{2}$ | | |
|---|---|---|---|---|---|---|---|
| $x_0$ | $x_1$ | $x_2$ | $x_3$ | $x_4$ | $x_5$ | rhs | |
| 1 | 0 | 0 | $\frac{7}{22}$ | $\frac{5}{22}$ | 0 | $\frac{21}{2}$ | ← |
| 0 | 1 | 0 | $\frac{3}{22}$ | $-\frac{1}{22}$ | 0 | $\frac{7}{2}$ | |
| 0 | 0 | 1 | $\frac{1}{22}$ | $\frac{7}{22}$ | 0 | $\frac{7}{2}$ | |
| 0 | 0 | 0 | $\frac{3}{2}$ | $\frac{5}{2}$ | 1 | $\frac{55}{2}$ | |

| $10\frac{1}{2}$ | $3\frac{1}{2}$ | $3\frac{1}{2}$ | $0$ | $0$ | $27\frac{1}{2}$ | $-\frac{1}{2}$ | |
|---|---|---|---|---|---|---|---|
| $x_0$ | $x_1$ | $x_2$ | $x_3$ | $x_4$ | $x_5$ | $x_6$ | rhs |
| 1 | 0 | 0 | $\frac{7}{22}$ | $\frac{5}{22}$ | 0 | 0 | $\frac{21}{2}$ |
| 0 | 1 | 0 | $\frac{3}{22}$ | $-\frac{1}{22}$ | 0 | 0 | $\frac{7}{2}$ |
| 0 | 0 | 1 | $\frac{1}{22}$ | $\frac{7}{22}$ | 0 | 0 | $\frac{7}{2}$ |
| 0 | 0 | 0 | $\frac{3}{2}$ | $\frac{5}{2}$ | 1 | 0 | $\frac{55}{2}$ |
| 0 | 0 | 0 | $-\frac{7}{22}$ | $[-\frac{5}{22}]$ | 0 | 1 | $-\frac{1}{2}$ |

| $10$ | $3\frac{3}{5}$ | $2\frac{4}{5}$ | $0$ | $2\frac{1}{5}$ | $22$ | $0$ | |
|---|---|---|---|---|---|---|---|
| $x_0$ | $x_1$ | $x_2$ | $x_3$ | $x_4$ | $x_5$ | $x_6$ | rhs |
| 1 | 0 | 0 | 0 | 0 | 0 | 1 | 10 |
| 0 | 1 | 0 | $\frac{1}{5}$ | 0 | 0 | $-\frac{1}{5}$ | $\frac{18}{5}$ ← |
| 0 | 0 | 1 | $-\frac{2}{5}$ | 0 | 0 | $\frac{7}{5}$ | $\frac{14}{5}$ |
| 0 | 0 | 0 | $-2$ | 0 | 1 | 11 | 22 |
| 0 | 0 | 0 | $\frac{7}{5}$ | 1 | 0 | $-\frac{22}{5}$ | $\frac{11}{5}$ |

| $10$ | $3\frac{3}{5}$ | $2\frac{4}{5}$ | $0$ | $2\frac{1}{5}$ | $22$ | $0$ | $-\frac{3}{5}$ | |
|---|---|---|---|---|---|---|---|---|
| $x_0$ | $x_1$ | $x_2$ | $x_3$ | $x_4$ | $x_5$ | $x_6$ | $x_7$ | rhs |
| 1 | 0 | 0 | 0 | 0 | 0 | 1 | 0 | 10 |
| 0 | 1 | 0 | $\frac{1}{5}$ | 0 | 0 | $-\frac{1}{5}$ | 0 | $\frac{18}{5}$ |
| 0 | 0 | 1 | $-\frac{2}{5}$ | 0 | 0 | $\frac{7}{5}$ | 0 | $\frac{14}{5}$ |
| 0 | 0 | 0 | $-2$ | 0 | 1 | 11 | 0 | 22 |
| 0 | 0 | 0 | $\frac{7}{5}$ | 1 | 0 | $-\frac{22}{5}$ | 0 | $\frac{11}{5}$ |
| 0 | 0 | 0 | $[-\frac{1}{5}]$ | 0 | 0 | $-\frac{4}{5}$ | 1 | $-\frac{3}{5}$ |

| 10 | 3 | 4 | 3 | −2 | 28 | 0 | 0 | |
|---|---|---|---|---|---|---|---|---|
| $x_0$ | $x_1$ | $x_2$ | $x_3$ | $x_4$ | $x_5$ | $x_6$ | $x_7$ | rhs |
| 1 | 0 | 0 | 0 | 0 | 0 | 1 | 0 | 10 |
| 0 | 1 | 0 | 0 | 0 | 0 | −1 | 1 | 3 |
| 0 | 0 | 1 | 0 | 0 | 0 | 3 | −2 | 4 |
| 0 | 0 | 0 | 0 | 0 | 1 | 19 | −10 | 28 |
| 0 | 0 | 0 | 0 | 1 | 0 | [−10] | 7 | −2 |
| 0 | 0 | 0 | 1 | 0 | 0 | 4 | −5 | 3 |

| $9\frac{4}{5}$ | $3\frac{1}{5}$ | $3\frac{2}{5}$ | $2\frac{1}{5}$ | 0 | $24\frac{1}{5}$ | $\frac{1}{5}$ | 0 | |
|---|---|---|---|---|---|---|---|---|
| $x_0$ | $x_1$ | $x_2$ | $x_3$ | $x_4$ | $x_5$ | $x_6$ | $x_7$ | rhs |
| 1 | 0 | 0 | 0 | $\frac{1}{10}$ | 0 | 0 | $\frac{7}{10}$ | $\frac{49}{5}$ ← |
| 0 | 1 | 0 | 0 | $-\frac{1}{10}$ | 0 | 0 | $\frac{3}{10}$ | $\frac{16}{5}$ |
| 0 | 0 | 1 | 0 | $\frac{3}{10}$ | 0 | 0 | $\frac{1}{10}$ | $\frac{17}{5}$ |
| 0 | 0 | 0 | 0 | $\frac{19}{10}$ | 1 | 0 | $\frac{33}{10}$ | $\frac{121}{5}$ |
| 0 | 0 | 0 | 0 | $-\frac{1}{10}$ | 0 | 1 | $-\frac{7}{10}$ | $\frac{1}{5}$ |
| 0 | 0 | 0 | 1 | $\frac{2}{5}$ | 0 | 0 | $-\frac{11}{5}$ | $\frac{11}{5}$ |

| $9\frac{4}{5}$ | $3\frac{1}{5}$ | $3\frac{2}{5}$ | $2\frac{1}{5}$ | $0$ | $24\frac{1}{5}$ | $\frac{1}{5}$ | $0$ | $-\frac{4}{5}$ | |
|---|---|---|---|---|---|---|---|---|---|
| $x_0$ | $x_1$ | $x_2$ | $x_3$ | $x_4$ | $x_5$ | $x_6$ | $x_7$ | $x_8$ | rhs |
| 1 | 0 | 0 | 0 | $\frac{1}{10}$ | 0 | 0 | $\frac{7}{10}$ | 0 | $\frac{49}{5}$ |
| 0 | 1 | 0 | 0 | $-\frac{1}{10}$ | 0 | 0 | $\frac{3}{10}$ | 0 | $\frac{16}{5}$ |
| 0 | 0 | 1 | 0 | $\frac{3}{10}$ | 0 | 0 | $\frac{1}{10}$ | 0 | $\frac{17}{5}$ |
| 0 | 0 | 0 | 0 | $\frac{19}{10}$ | 1 | 0 | $\frac{33}{10}$ | 0 | $\frac{121}{5}$ |
| 0 | 0 | 0 | 0 | $-\frac{1}{10}$ | 0 | 1 | $-\frac{7}{10}$ | 0 | $\frac{1}{5}$ |
| 0 | 0 | 0 | 1 | $\frac{2}{5}$ | 0 | 0 | $-\frac{11}{5}$ | 0 | $\frac{11}{5}$ |
| 0 | 0 | 0 | 0 | $[-\frac{1}{10}]$ | 0 | 0 | $-\frac{7}{10}$ | 1 | $-\frac{4}{5}$ |

| 9 | 4 | 1 | −1 | 8 | 9 | 1 | 0 | 0 | |
|---|---|---|---|---|---|---|---|---|---|
| $x_0$ | $x_1$ | $x_2$ | $x_3$ | $x_4$ | $x_5$ | $x_6$ | $x_7$ | $x_8$ | rhs |
| 1 | 0 | 0 | 0 | 0 | 0 | 0 | 0 | 1 | 9 |
| 0 | 1 | 0 | 0 | 0 | 0 | 0 | 1 | −1 | 4 |
| 0 | 0 | 1 | 0 | 0 | 0 | 0 | −2 | 3 | 1 |
| 0 | 0 | 0 | 0 | 0 | 1 | 0 | −10 | 19 | 9 |
| 0 | 0 | 0 | 0 | 0 | 0 | 1 | 0 | −1 | 1 |
| 0 | 0 | 0 | 1 | 0 | 0 | 0 | [−5] | 4 | −1 |
| 0 | 0 | 0 | 0 | 1 | 0 | 0 | 7 | −10 | 8 |

| | | | | | | | | | |
|---|---|---|---|---|---|---|---|---|---|
| 9 | $3\frac{4}{5}$ | $1\frac{2}{5}$ | 0 | $6\frac{3}{5}$ | 11 | 1 | $\frac{1}{5}$ | 0 | |
| $x_0$ | $x_1$ | $x_2$ | $x_3$ | $x_4$ | $x_5$ | $x_6$ | $x_7$ | $x_8$ | rhs |
| 1 | 0 | 0 | 0 | 0 | 0 | 0 | 0 | 1 | 9 |
| 0 | 1 | 0 | $\frac{1}{5}$ | 0 | 0 | 0 | 0 | $-\frac{1}{5}$ | $\frac{19}{5}$ ← |
| 0 | 0 | 1 | $-\frac{2}{5}$ | 0 | 0 | 0 | 0 | $\frac{7}{5}$ | $\frac{7}{5}$ |
| 0 | 0 | 0 | $-2$ | 0 | 1 | 0 | 0 | 11 | 11 |
| 0 | 0 | 0 | 0 | 0 | 0 | 1 | 0 | $-1$ | 1 |
| 0 | 0 | 0 | $-\frac{1}{5}$ | 0 | 0 | 0 | 1 | $-\frac{4}{5}$ | $\frac{1}{5}$ |
| 0 | 0 | 0 | $\frac{7}{5}$ | 1 | 0 | 0 | 0 | $-\frac{22}{5}$ | $\frac{33}{5}$ |

| | | | | | | | | | | |
|---|---|---|---|---|---|---|---|---|---|---|
| 9 | $3\frac{4}{5}$ | $1\frac{2}{5}$ | 0 | $6\frac{3}{5}$ | 11 | 1 | $\frac{1}{5}$ | 0 | $-\frac{4}{5}$ | |
| $x_0$ | $x_1$ | $x_2$ | $x_3$ | $x_4$ | $x_5$ | $x_6$ | $x_7$ | $x_8$ | $x_9$ | rhs |
| 1 | 0 | 0 | 0 | 0 | 0 | 0 | 0 | 1 | 0 | 9 |
| 0 | 1 | 0 | $\frac{1}{5}$ | 0 | 0 | 0 | 0 | $-\frac{1}{5}$ | 0 | $\frac{19}{5}$ |
| 0 | 0 | 1 | $-\frac{2}{5}$ | 0 | 0 | 0 | 0 | $\frac{7}{5}$ | 0 | $\frac{7}{5}$ |
| 0 | 0 | 0 | $-2$ | 0 | 1 | 0 | 0 | 11 | 0 | 11 |
| 0 | 0 | 0 | 0 | 0 | 0 | 1 | 0 | $-1$ | 0 | 1 |
| 0 | 0 | 0 | $-\frac{1}{5}$ | 0 | 0 | 0 | 1 | $-\frac{4}{5}$ | 0 | $\frac{1}{5}$ |
| 0 | 0 | 0 | $\frac{7}{5}$ | 1 | 0 | 0 | 0 | $-\frac{22}{5}$ | 0 | $\frac{33}{5}$ |
| 0 | 0 | 0 | $[-\frac{1}{5}]$ | 0 | 0 | 0 | 0 | $-\frac{4}{5}$ | 1 | $-\frac{4}{5}$ |

| 9 | 3 | 3 | 4 | 1 | 19 | 1 | 1 | 0 | 0 | |
|---|---|---|---|---|---|---|---|---|---|---|
| $x_0$ | $x_1$ | $x_2$ | $x_3$ | $x_4$ | $x_5$ | $x_6$ | $x_7$ | $x_8$ | $x_9$ | rhs |
| 1 | 0 | 0 | 0 | 0 | 0 | 0 | 0 | 1 | 0 | 9 |
| 0 | 1 | 0 | 0 | 0 | 0 | 0 | 0 | −1 | 1 | 3 |
| 0 | 0 | 1 | 0 | 0 | 0 | 0 | 0 | 3 | −2 | 3 |
| 0 | 0 | 0 | 0 | 0 | 1 | 0 | 0 | 19 | −10 | 19 |
| 0 | 0 | 0 | 0 | 0 | 0 | 1 | 0 | −1 | 0 | 1 |
| 0 | 0 | 0 | 0 | 0 | 0 | 0 | 1 | 0 | −1 | 1 |
| 0 | 0 | 0 | 0 | 1 | 0 | 0 | 0 | −10 | 7 | 1 |
| 0 | 0 | 0 | 1 | 0 | 0 | 0 | 0 | 4 | −5 | 4 |

In the space of the original variables, the sequence of Gomory cutting-planes is

(*A*) $2x_1 + x_2 \leq 10;$

(*B*) $3x_1 + x_2 \leq 13;$

(*C*) $2x_1 + x_2 \leq 9;$

(*D*) $3x_1 + x_2 \leq 12,$

and the sequence of optimal linear-programming solutions (in the original variables) is:

(*a*) $x_1 = 3\frac{1}{2}, \quad x_2 = 3\frac{1}{2};$

(*b*) $x_1 = 3\frac{3}{5}, \quad x_2 = 2\frac{4}{5};$

(*c*) $x_1 = 3\frac{1}{5}, \quad x_2 = 3\frac{2}{5};$

(*d*) $x_1 = 3\frac{4}{5}, \quad x_2 = 1\frac{2}{5};$

(*e*) $x_1 = 3, \quad x_2 = 3.$

Toward describing a way to guarantee that Gomory's Cutting-Plane Method will terminate in a finite number of steps, we need to understand something about unbounded integer programs.

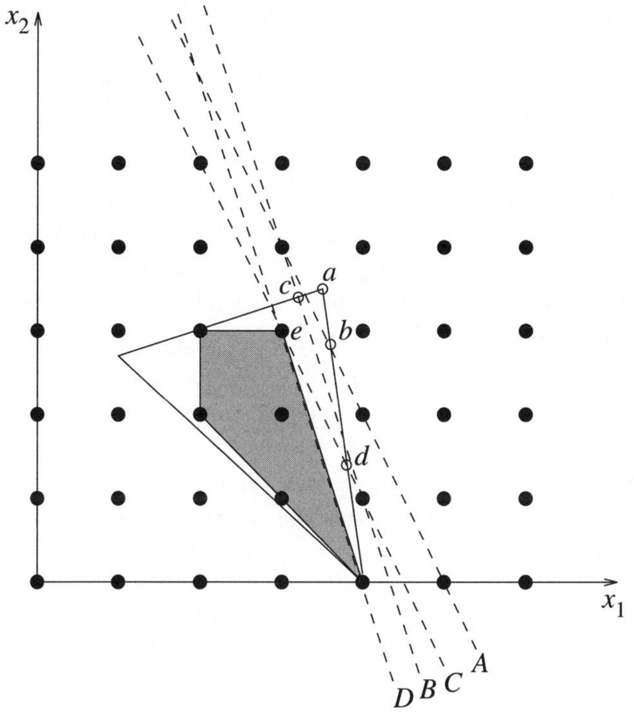


♠

**Problem (Unbounded integer program).** Assume that the data for *IP* are rational and that *IP* has a feasible solution. Prove that the linear-programming relaxation of *IP* is unbounded if and only if *IP* is unbounded.

We assume that *IP* has an optimal solution. With this assumption, the linear-programming relaxation of *IP* cannot have unbounded objective value. Therefore, we can append a redundant constraint to the formulation and produce a dual-feasible linear-programming basis. Under these assumptions, Gomory demonstrated that there is a refinement of his cutting-plane method that is guaranteed to terminate.

**Theorem (Finiteness of Gomory's Cutting-Plane Method).** *Assume that IP has an optimal solution. Gomory's Cutting-Plane Method terminates provided that*

a. *the source row is chosen to be the one with the noninteger basic variable of least index,*

b. *the slack variable in each appended equation is given the next available index,*
c. *the Epsilon-Perturbed Dual Simplex Method is used to reoptimize after each equation is appended.*

*Proof.* When we append a Gomory cutting plane in the form of $\overline{G}_{\beta_i}$ to an optimal simplex table, the value of the new basic variable for that row is negative. The objective coefficient that we assign to the slack variable $x_k$ is $\epsilon^k$, as per the Epsilon-Perturbed Dual Simplex Method. Therefore, the objective value for the perturbed problem, even before we pivot, changes by $\left(\lfloor x^*_{\beta_i} \rfloor - x^*_{\beta_i}\right) \epsilon^k$, which is a decrease. Then we reoptimize with the dual simplex method, and the objective value for the perturbed problem continues to decrease, at each iteration of the dual simplex method.

Consider the first iteration of the dual simplex method, just after $\overline{G}_{\beta_i}$ is appended to the simplex table. Some nonbasic variable, say $x_{\eta_j}$, is exchanged with the basic variable $x_k$. The value of $x_{\beta_i}$ changes from $x^*_{\beta_i}$ to

$$\widetilde{x}^*_{\beta_i} := x^*_{\beta_i} - \overline{a}_{\beta_i \eta_j} \frac{x^*_{\beta_i} - \lfloor x^*_{\beta_i} \rfloor}{a_{\beta_i \eta_j} - \lfloor a_{\beta_i \eta_j} \rfloor} .$$

Because

$$\frac{x^*_{\beta_i} - \lfloor x^*_{\beta_i} \rfloor}{a_{\beta_i \eta_j} - \lfloor a_{\beta_i \eta_j} \rfloor}$$

is positive, a decrease in $x_{\beta_i}$ means that $\overline{a}_{\beta_i \eta_j} > 0$. Therefore,

$$\frac{\overline{a}_{\beta_i \eta_j}}{a_{\beta_i \eta_j} - \lfloor a_{\beta_i \eta_j} \rfloor} \geq 1.$$

We conclude that $\widetilde{x}^*_{\beta_i} \leq \lfloor x^*_{\beta_i} \rfloor$

Assume that the optimal objective value of the linear-programming relaxation of *IP* is $z^*$. Let $x^{LP}$ be the optimal solution the linear-programming relaxation of *IP* found by the Epsilon-Perturbed Dual Simplex Method.

Consider the box of lattice points

$$B := \left\{ x \in \mathbf{Z}^{n+1} \ : \ z^* \leq x_0 \leq x_0^{LP};\ 0 \leq x_j \leq x_j^{LP},\ \text{for } j = 1, 2, \ldots, n \right\}.$$

The finite set $B$ can be ordered *lexicographically*: $x^1 \prec x^2$ if $\sum_{j=0}^{n} x^1_j \epsilon^j < \sum_{j=0}^{n} x^2_j \epsilon^j$ for arbitrarily small positive $\epsilon$. After each successive reoptimization (by the Epsilon-Perturbed Dual Simplex Method) in which a source row with an original variable (i.e., $x_0, x_1, \ldots, x_n$) is selected, the solution has the values of its original variables lexicographically less than an element of $B$ that is successively lexicographically smaller. Therefore, after a finite number of iterations, all original variables take on integer values. ∎

## 6.4 Tightening a Constraint

Let $a_{ij}$ $(1 \le i \le m,\ 1 \le j \le n)$ and $b_i$ $(1 \le i \le m)$ be integers, and consider the integer linear program

$$
(IP)\qquad
\begin{aligned}
\max \sum_{j=1}^{n} c_j x_j \\
\text{subject to:} \\
\sum_{j=1}^{n} a_{ij} x_j \le b_i, &\quad \text{for } i = 1, 2, \ldots, m; \\
0 \le x_j \le 1, &\quad \text{for } j = 1, 2, \ldots, n; \\
x_j \in \mathbf{Z}, &\quad \text{for } j = 1, 2, \ldots, n.
\end{aligned}
$$

There are several strategies, based on a single constraint, that can be used to strengthen *IP*. Consider a single inequality

$$\sum_{j=1}^{n} a_{kj} x_j \le b_k.$$

As we consider only one constraint at a time, we can assume that $a_{kj} \ge 0$ for all $j = 1, 2, \ldots, n$ (we can substitute $1 - x'_j$ for $x_j$ if $a_{kj} < 0$).

Provided that $b_k$ is not too large, we can use a recursion to efficiently solve the knapsack program (see the "Knapsack program" Problem)

$$
(KP_k)\qquad
\begin{aligned}
b'_k := \max \sum_{j=1}^{n} a_{kj} x_j \\
\text{subject to:} \\
\sum_{j=1}^{n} a_{kj} x_j \le b_k; \\
0 \le x_j \le 1, &\quad \text{for } j = 1, 2, \ldots, n; \\
x_j \in \mathbf{Z}, &\quad \text{for } j = 1, 2, \ldots, n,
\end{aligned}
$$

and then replace $b_k$ with $b'_k$ in *IP*.

Another strategy is to attempt to find sets $W \subset \{1, 2, \ldots, n\}$ satisfying

$$\sum_{j \in W} a_{kj} > b_k.$$

Then the *cover inequality*

$$\sum_{j \in W} x_j \leq |W| - 1$$

is satisfied by all feasible solutions of *IP*. The cover inequality is *minimal* if

$$\sum_{j \in W - l} a_{kj} \leq b_k, \quad \forall\, l \in W.$$

We can generate violated cover inequalities for use in a cutting-plane algorithm. For fixed $x^*$, let

$$(DP_k) \qquad \begin{aligned} \sigma(x^*) := \min \sum_{j=1}^{n} (1 - x_j^*) z_j \\ \sum_{j=1}^{n} a_{kj} z_j > b_k; \\ z_j \in \{0, 1\}, \text{ for } j = 1, 2, \ldots n. \end{aligned}$$

**Theorem (Finding violated cover inequalities).** *If $\sigma(x^*) \geq 1$, then all cover inequalities for the constraint $\sum_{j=1}^{n} a_{kj} x_j \leq b_k$ are satisfied by $x^*$. Alternatively, if $\sigma(x^*) < 1$ and $z^*$ solves $DP_k$, then $W := S(z^*)$ describes a cover inequality that is violated by $x^*$.*

*Proof.* Notice that $z$ is a feasible solution of $DP_k$ if and only if $z$ is the characteristic vector of a set $W$ that describes a cover inequality. If $\sigma(x^*) \geq 1$, then $\sum_{j=1}^{n} (1 - x_j^*) z_j \geq 1$ for all $z$ that are feasible for $DP_k$. Therefore, $\sum_{j=1}^{n} x_j^* z_j \leq \sum_{j=1}^{n} z_j - 1$ for all $z$ that are feasible for $DP_k$. That is, all cover inequalities are satisfied by $x^*$. Alternatively, if $\sigma(x^*) < 1$, then $\sum_{j=1}^{n} (1 - x_j^*) z_j^* < 1$ for some $z^*$ that is feasible to $DP_k$. It follows that $|S(z^*)| - \sum_{j \in S(z^*)} x_j^* < 1$, which implies that the valid cover inequality $\sum_{j \in S(z^*)} x_j \leq |S(z^*)| - 1$ is violated by $x^*$. ∎

**Problem (Cover separation).** Describe how $DP_k$ can be solved by use of knapsack-program methods (see the "Knapsack program" Problem).

Still another strategy is to attempt to find sets $W \subset \{1, 2, \ldots, n\}$ satisfying

$$a_{kj} + a_{kl} > b_k, \quad \forall \text{ distinct } j, l \in W.$$

Then the clique inequality

$$\sum_{j \in W} x_j \leq 1$$

is satisfied by all feasible solutions of *IP*. The clique inequality is *maximal* if for all $i \notin W$,

$$a_{kj} + a_{ki} \leq b_k, \quad \text{for some } j \in W.$$

> **Problem (Clique separation).** Describe how the separation problem for clique inequalities can be recast as a maximum-weight vertex-packing problem.

Suppose that we have the valid inequality $\sum_{j=1}^{n} \alpha_j x_j \leq \beta$. We may assume that all $\alpha_j$ are nonnegative (by complementing variables if necessary). We can choose an index $k$ and consider the *lifting program*

$$\delta_k := \beta - \max \sum_{j \neq k} \alpha_j x_j$$

subject to:

$$\sum_{j \neq k} a_{ij} x_j \leq b_i - a_{ik}, \quad \text{for } i = 1, 2, \ldots, m;$$
$$0 \leq x_j \leq 1\,, \quad \text{for } j \neq k;$$
$$x_j \in \mathbf{Z}\,, \quad \text{for } j \neq k.$$

If the lifting program is infeasible, then $x_k = 0$ is valid. Otherwise, the inequality $\alpha'_k x_k + \sum_{j \neq k} \alpha_j x_j \leq \beta$ is valid, for all $\alpha'_k \leq \delta_k$. In practice, it may not be practical to compute more than a lower bound on $\delta_k$ (which amounts to computing an upper bound on the maximum in the lifting program). Lifting can be applied to any valid inequality, and, sequentially, to each variable.

Consider the formulation

$$7x_1 + 2x_2 + x_3 + 3x_4 + 6x_5 + 5x_6 \leq 8;$$
$$0 \leq x_1, x_2, x_3, x_4, x_5, x_6 \leq 1, \text{ integer.}$$

Let $\mathcal{P}$ be the convex hull of the feasible solutions. Because the 0-vector and the six standard unit vectors are feasible, $\dim(\mathcal{P}) = 6$. It turns out that $\mathcal{P}$ has 18 integer-valued points,

$$\begin{gathered}(000000),\\(100000),\\(010000),\\(001000),\\(000100),\\(000010),\\(000001),\\(101000),\\(011000),\\(010100),\\(010010),\\(010001),\\(001100),\\(001010),\\(001001),\\(000101),\\(011100),\\(011001),\end{gathered}$$

and 14 facets, described by the following inequalities:

(1) $x_1 \geq 0;$

(2) $x_2 \geq 0;$

(3) $x_3 \geq 0;$

(4) $x_4 \geq 0;$

(5) $x_5 \geq 0;$

(6) $x_6 \geq 0;$

(7) $x_3 \leq 1;$

(8) $x_1 + x_2 \leq 1;$

(9) $x_1 + x_4 + x_5 \leq 1;$

(10) $x_1 + x_5 + x_6 \leq 1;$

(11) $x_1 + x_2 + x_3 + x_5 \leq 2;$

(12) $x_1 + x_3 + x_4 + x_5 + x_6 \leq 2;$

(13) $2x_1 + x_2 + x_4 + x_5 + x_6 \leq 2;$

(14) $3x_1 + x_2 + x_3 + 2x_4 + 3x_5 + 2x_6 \leq 4.$

Inequalities (1)–(7) come from the initial formulation. (8) can be thought of as a maximal clique inequality or a minimal cover inequality. (9) and (10) are maximal clique inequalities. We can realize (11) by starting with the minimal cover inequality

$$x_2 + x_3 + x_5 \leq 2,$$

and then lifting the coefficient of $x_1$. We can obtain (12) by starting with the minimal cover inequality

$$x_3 + x_4 + x_6 \leq 2,$$

and then, sequentially, lifting the coefficients of $x_1$ and $x_4$. We can obtain (13) by starting with the minimal cover inequality

$$x_2 + x_4 + x_6 \leq 2,$$

and, sequentially, lifting the coefficients of $x_1$ and $x_5$. One way to see how (14) can arise is to add (12) and (13) and then lift the coefficient of $x_5$.

## 6.5 Constraint Generation for Combinatorial-Optimization Problems

Some integer programs require an enormous number of constraints, relative to the number of "natural" variables. For example, let $G$ be a simple graph, and let $\mathcal{P}(G)$ denote the convex hull of the characteristic vectors of Hamiltonian tours of $G$. We can formulate the problem of finding a minimum-weight Hamiltonian tour:

$$\min \sum_{e \in E(G)} c(e)\, x_e$$

subject to:

$$\sum_{e \in \delta_G(v)} x_e = 2, \quad \forall\, v \in V(G) \quad \textit{(degree constraints)};$$

$$\sum_{e \in E(G[W])} x_e \leq |W| - 1, \quad \begin{array}{l} \forall\, W \subset V(G): \\ 3 \leq |W| \leq |V(G)| - 3 \end{array} \quad \textit{(subtour-elimination inequalities)};$$

$$0 \leq x_e \leq 1, \quad \forall\, e \in E(G) \quad \textit{(simple lower- and upper-bound inequalities)};$$

$$x_e \in \mathbf{Z}, \quad \forall\, e \in E(G).$$

It is usually impractical to explicitly list all of the subtour-elimination inequalities. However, we can treat them just like cutting planes and generate them as needed; all that is required is an efficient algorithm for generating a single violated inequality with respect to a point $x^* \in \mathbf{R}_+^{E(G)}$.

First, we note that either $W := S$ or $W := V(G) \setminus S$ describes a subtour-elimination inequality that is violated by $x^*$ when

$$\sum_{e \in \delta_G(S)} x_e^* < 2.$$

This easily follows from $\sum_{e \in E(G)} x_e = |V(G)|$ (which follows from the degree constraints). Conversely, if

$$\sum_{e \in \delta_G(S)} x_e^* \geq 2$$

for all $S \subset V(G)$, then $x^*$ satisfies all subtour-elimination inequalities.

**Separation Algorithm for Subtour-Elimination Inequalities**

1. Form a digraph $G'$ with $V(G') := V(G)$. $E(G')$ is obtained by the replacement of each edge $e \in E(G)$ with a directed edge $\overline{e}$ in $G'$ (in either direction).
2. Define an upper-bound function $c : E(G') \mapsto \mathbf{R}$ by $c(\overline{e}) := x_e^*$ and a lower-bound function $l : E(G') \mapsto \mathbf{R}$ by $l(\overline{e}) := -x_e^*$.
3. Distinguish any vertex $v$, and calculate minimum-capacity $v$–$w$ cutsets $S_w$ for all $w \in V(G') - v$.
4. Let $S$ be any choice of $S_w$ $(w \in V(G') - v)$ so that $C(S) = \min C(S_w)$.
   i. If $C(S) \geq 2$, then all subtour-elimination inequalities are satisfied by $x^*$.
   ii. If $C(S) < 2$, then either $W := S$ or $W := V(G) \setminus S$ describes a subtour-elimination inequality that is violated by $x^*$.

**Problem (Prize-collecting traveling salesperson).** Let $G$ be an undirected graph with a selected "starting vertex" $v \in V(G)$. We have positive weight functions $f$ on $V(G) - v$ and $c$ on $E(G)$. The cost of traveling along edge $\{i, j\} \in E(G)$ is $c(e)$. The revenue obtained for visiting vertex $i$ is $f(i)$. Starting from $v$, we must return to $v$, and we must visit other vertices no more than once each. We want to find the trip with maximum net profit. Explain why the following formulation of the problem is correct:

$$\max \sum_{i \in V(G)-v} f(i) y_i - \sum_{e \in E(G)} c(e) x_e$$

subject to:

$$\sum_{e \in \delta_G(i)} x_e = 2y_i, \quad \forall\, i \in V(G);$$

$$(*) \qquad \sum_{e \in E(G[W])} x_e \le \sum_{i \in W-w} y_i, \quad \forall\, W \subset V(G),\ w \in W;$$

$$0 \le x_e \le 1, \quad \forall\, e \in E(G) \setminus \delta_G(v);$$

$$0 \le x_e \le 2, \quad \forall\, e \in \delta_G(v);$$

$$0 \le y_i \le 1, \quad \forall\, i \in V(G) - v;$$

$$x_e \in \mathbf{Z}, \quad \forall\, e \in E(G);$$

$$y_i \in \mathbf{Z}, \quad \forall\, i \in V(G) - v;$$

$$y_v = 1.$$

Now, suppose that $x^* \in \mathbf{R}_+^{E(G)}$, $y^* \in \mathbf{R}_+^{V(G)}$, and $w \in V(G)$ are fixed, and consider the *linear* program:

$$\max \sum_{e \in E(G) \setminus \delta_G(v)} x_e^* z_e - \sum_{i \in V(G) \setminus \{v, w\}} y_i^* u_i$$

subject to:

$$z_e \le u_i \text{ and } z_e \le u_j, \quad \forall\, \{i, j\} \in E(G) \setminus \delta_G(v);$$

$$0 \le z_e \le 1, \quad \forall\, e \in E(G) \setminus \delta_G(v);$$

$$0 \le u_i \le 1, \quad \forall\, i \in V(G) \setminus \{v, w\};$$

$$u_w = 1.$$

Prove that the solution of this linear program solves the separation problem for the inequalities $(*)$.

Even with all of the subtour-elimination inequalities, we do not have a complete description of the Hamiltonian-tour polytope $\mathcal{P}(G)$. However, using the Chvátal–Gomory process, we can derive valid inequalities that cut off some fractional solutions. An *elementary comb* of $G$ has a *handle* $G[W]$ and *teeth* $F \subset E(G)$, satisfying the following properties:

1. $3 \le |W| \le |V(G)| - 1$;
2. odd $|F| \ge 3$;
3. $F \subset \delta_G(W)$;
4. $F$ is a matching.

**Example (Comb).** Below is a depiction of an elementary comb with five teeth.

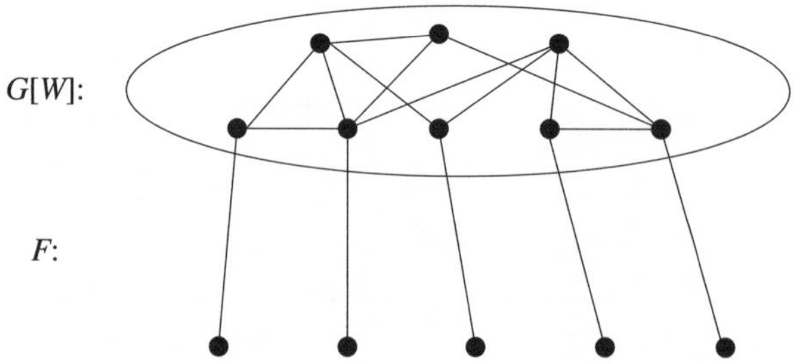


♠

The elementary comb determines the *2-factor inequality*

$$\sum_{e \in E(G[W])} x_e + \sum_{e \in F} x_e \le |W| + \left\lfloor \frac{|F|}{2} \right\rfloor.$$

**Theorem (Validity of 2-factor inequalities).** *If $S \subset E(G)$ satisfies $S \cap \delta_G(v) = 2$ for all $v \in V(G)$, then $x(S)$ satisfies the 2-factor inequalities.*

*Proof.* If $S \cap E(G[W])$ is a tour of $W$, then $W \cap F = \emptyset$. In that case, plugging $x(S)$ into the 2-factor inequality, we get $|W|$ on the left-hand side and at least that on the right-hand side.

If $S \cap E(G[W])$ is not a tour of $W$, then it consists of say, $p$ paths (some of which might consist of a single vertex). Then we will have $|S \cap E(G[W])| = |W| - p$. In that case, considering the degree constraints for vertices in $W$, we must have $|S \cap F| = 2p$. Then, plugging $x(S)$ into the 2-factor inequality, we get $|W| + p$ on the left-hand side and $|W| + \lfloor |F|/2 \rfloor$ on the right-hand side. Validity follows from the observation that $2p = |S \cap F| \le |F|$. ∎

**Problem (2-factor inequalities and Chvátal–Gomory cutting planes).** Demonstrate how 2-factor inequalities are Chvátal–Gomory cutting planes with respect to the degree constraints and the simple lower- and upper-bound inequalities.

**Exercise (Violated 2-factor inequality).** The following graph describes a fractional solution to the linear-programming relaxation of the integer-programming formulation for the maximum-weight Hamiltonian tour problem. The values of the variables are indicated on the edges. Convince yourself that this point satisfies the degree constraints and the subtour-elimination inequalities, and find a 2-factor inequality that it violates.

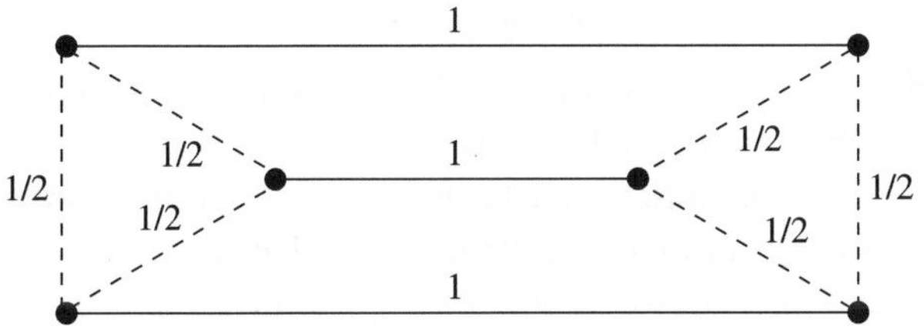


Even with all of the 2-factor inequalities, we do not have a complete description of $\mathcal{P}(G)$ (although the 2-factor inequalities, degree constraints, and simple lower- and upper-bound inequalities are enough to describe the convex hull of the characteristic vectors of the 2-factors of $G$). Many other families of valid inequalities are known for $\mathcal{P}(G)$, but it is unlikely that anyone will ever establish a satisfying linear-inequality description of $\mathcal{P}(G)$.

**Problem (Odd-cycle separation).** Let $H$ be an undirected graph with weight function $d$ on $E(H)$. Recall the minimum-weight cut problem (see p. 147)

$$\min_{S \subset V(H)} \left\{ \sum_{e \in \delta_H(S)} d(e) \right\} .$$

a. Explain why the following formulation is valid for the minimum-weight cut problem:

$$\min \sum_{e \in E(G)} d(e)\, x_e$$

subject to:

$$\sum_{e \in C} x_e \le |C| - 1, \quad \forall \text{ odd cycles } C \text{ of } H \quad (\textit{odd-cycle inequalities});$$

$$0 \le x_e \le 1, \quad \forall\, e \in E(G) \quad (\textit{simple lower- and upper-bound inequalities});$$

$$x_e \in \mathbf{Z}, \quad \forall\, e \in E(G).$$

b. Give an efficient algorithm for solving the separation problem for the odd-cycle inequalities. *Hint:* See the "Minimum-weight even path" Problem (p. 126).

## 6.6 Further Study

The work of Nemhauser and Wolsey (1988) is a useful starting point for more material on cutting-plane methods. Also, the book by Schrijver (1986) is an excellent resource for more material concerning integer linear programming.

There is a wealth of mathematical material and great computational success on cutting-plane methods for the minimum-weight Hamiltonian tour problem; see Applegate, Bixby, Chvátal and Cook (1998), and the references therein.

For the minimum-weight cut problem in undirected graphs, it turns out that the linear-programming relaxation of the formulation (given in part a of the "Odd-cycle separation" Problem, p. 175) solves the problem for planar graphs and also for bipartite graphs (see Chapter 75 of Schrijver (2003)).

# 7

# *Branch-&-Bound*

For general combinatorial-optimization problems, we do not know theoretically efficient algorithms. Indeed, it is very likely that theoretically efficient algorithms do not exist for some of our favorite combinatorial-optimization problems (e.g., the Metric Traveling-Salesperson's Problem, the Maximum-Cardinality Vertex-Packing Problem, and many other hard problems with compact descriptions as integer linear programs).

*Branch-&-Bound* is a "divide-and-conquer" framework for solving discrete-optimization problems. Because it is semienumerative and performs very badly in the worst case, it is not something that most people who work in discrete optimization are particularly proud of. Still, it is a very important part of the discrete/combinatorial-optimization tool kit – something like a "plumber's helper." The methodology is rather robust, it can often be integrated with other techniques (like cutting-plane techniques for integer linear programming), and, in many situations, it is partly responsible for the success in solving large instances of difficult problems.

We assume that our discrete-optimization problem has the general form

$$(P) \qquad z_P := \max\{c(S) \ : \ S \in \mathcal{S}\},$$

where $\mathcal{S}$ is a finite set and $c$ is an arbitrary function from $\mathcal{S}$ to $\mathbf{R}$. The Branch-&-Bound framework is based on three main ingredients:

1. *Upper Bounds*: Efficient methods for determining a good upper bound $UB(P)$ on $z_P$.
2. *Branching Rules*: Methods for replacing an instance $P$ of the discrete-optimization problem with some further "smaller" subproblem instances $P_\ell$, such that some optimal solution of $P$ maps to an optimal solution of a subproblem $P_\ell$.

3. *Lower Bounds*: Efficient heuristics that attempt to determine a feasible *candidate* solution $S \in \mathcal{S}$ with as high a value of $c(S)$ as is practical, yielding the lower bound $LB(P) := c(S)$ on $z_P$.

The algorithm maintains a global lower bound $LB$ on $z_P$ (see ingredient 3), and a list $\mathcal{L}$ of active subproblems with a subproblem upper bound (see ingredient 1) for each. Initially, we set $\mathcal{L} := \{P\}$ and we calculate $UB(P)$. Initially, we apply the lower-bounding heuristic to $P$ and set $LB := LB(P)$. If the lower-bounding heuristic fails to provide a feasible solution, then we initially set $LB := -\infty$.

At a general step, we remove a subproblem $P'$ from the list $\mathcal{L}$. If its subproblem upper bound $UB(P')$ is less than or equal to the global lower bound $LB$, then we discard $P'$ – in this case, we say that $P'$ is *fathomed by bounds*. If, alternatively, the subproblem upper bound $UB(P')$ is greater than the global lower bound $LB$, we create further subproblem "children" of $P'$, say $P'_\ell$, according to the branching rule (see ingredient 2). For each subproblem $P'_\ell$ that we create, we compute its subproblem upper bound $UB(P'_\ell)$ and possibly a subproblem lower bound. If the subproblem upper bound is less than or equal to the global lower bound $LB$, then we discard $P'_\ell$ (fathoming by bounds, as above). If we have some logic for determining, in the subproblem upper-bounding procedure, that $P'_\ell$ is infeasible, then we consider the subproblem upper bound for $P'_\ell$ to be $-\infty$, and it will also be discarded – in this case, we say that $P'_\ell$ has been *fathomed by infeasibility*. If we obtain a finite subproblem lower bound $LB(P'_\ell)$ for $P'_\ell$, then we update the global lower bound: $LB \leftarrow \max\{LB, LB(P'_\ell)\}$, and then we discard $P'_\ell$ if its subproblem lower bound $LB(P'_\ell)$ is equal to it subproblem upper bound – in this case, we say that $P'_\ell$ is *fathomed by optimality*.

If a subproblem child $P'_\ell$ is not fathomed according to the preceding possibilities, then we put $P'_\ell$ on the list $\mathcal{L}$. If the list $\mathcal{L}$ is empty after we process all of the subproblem children $P'_\ell$ of $P'$, then we stop with the conclusion that $z_P = LB$. Otherwise, we remove another subproblem from the list and repeat all of the preceding procedure.

Finite termination is ensured by having each subproblem instance be "smaller" than its parent (see ingredient 2). The exact meaning of "smaller" depends on the application; this will be developed further in the remaining sections of this chapter.

Although not necessary for carrying out the algorithm, some additional information can be recorded that can be useful. Just before removing a subproblem from $\mathcal{L}$ to process, we may calculate the *global upper bound*

$$UB := \max\{LB, \max\{UB(P') \ : \ P' \in \mathcal{L}\}\}.$$

At the beginning of the execution of the algorithm, we will have $UB := UB(P)$. At any stage, because $LB \le z_P \le UB$, we may stop the algorithm when $UB$ is deemed to be close enough to *LB*, at which point we know that the objective value of the candidate solution is within $UB - LB$ of the optimal objective value $z_P$. In addition, the global upper bound can be used to develop a useful branching rule (to be discussed shortly).

The Branch-&-Bound framework can be effective only if, after we remove a subproblem $P'$ from the list $\mathcal{L}$, we do not replace it on the list with very many of its children $P'_\ell$ very often. Our success, in this regard, depends on the quality of the lower and upper bounds.

There is always the flexibility of being able to choose any subproblem $P'$ from $\mathcal{L}$ to remove. Experience indicates that a "Last-in/First-out" discipline may obtain a good feasible solution relatively quickly (assuming that the lower-bounding heuristic is pretty good), and so this discipline is aimed at increasing the global lower bound *LB*. On the other hand, the "Best-Bound" discipline of choosing a subproblem $P'$ with the property that $UB(P') = UB$ has the goal of trying to decrease the global upper bound *UB* relatively quickly. A prudent strategy seems to be to use Last-in/First-out early in the process and then employ mostly Best-Bound, reverting to Last-in/First-out for some time if $\mathcal{L}$ grows large or if the global upper bound has not decreased significantly after many iterations.

In the remainder of this chapter, we will see how the Branch-&-Bound framework can be used for particular problems.

## 7.1 Branch-&-Bound Using Linear-Programming Relaxation

Consider the integer linear program

$$(P) \qquad \begin{aligned} z_P := \max \sum_{j=1}^{n} c_j x_j \\ \text{subject to:} \\ \sum_{j=1}^{n} a_{ij} x_j \le b_i, \quad \text{for } i = 1, 2, \ldots, m; \\ x_j \in \mathbf{Z}, \quad \text{for } j = 1, 2, \ldots, n. \end{aligned}$$

For simplicity, we assume that the feasible region of the linear-programming relaxation of $P$ is a bounded set. Therefore, the set of feasible solutions to any subproblem $P'$ is a finite set.

1. *Upper Bounds*: We solve the linear-programming relaxation of a subproblem $P'$. Let $x^*$ be its optimal solution. The optimal objective value of the linear-programming relaxation is an upper bound on $z_{P'}$.
2. *Branching Rule*: We choose a variable $x_k$ for which $x_k^* \notin \mathbf{Z}$. We branch by creating two new subproblems: (a) $P'$ together with the additional inequality $x_k \le \lfloor x_k^* \rfloor$, and (b) $P'$ together with the additional inequality $x_k \ge \lceil x_k^* \rceil$.

3. *Lower Bounds*: If the solution $x^*$ of the linear-programming relaxation of $P'$ happens to be in $\mathcal{S}'$ (the set of feasible solutions to the integer linear program $P'$), then its objective $z^* := c(x^*)$ value is a lower bound on $z_{P'}$. We can make this a bit more sophisticated. Even if the optimal solution is not feasible to $P'$, we may visit feasible solutions to $P'$ in the process of solving its linear-programming relaxation (if, for example, we solve the subproblems by the primal simplex method). Any of these also provides a lower bound on $z_{P'}$. Finally, it may be possible to perturb the solution of the linear-programming relaxation to obtain a feasible solution of $P'$; for example, if the $a_{ij}$ are nonnegative, then rounding the components of $x^*$ down provides a feasible solution.

This Branch-&-Bound method can be effective only if we do not replace one subprogram on the list with two very often. This depends on the quality of the bounds we obtain by solving the linear-programming relaxations. Often, these bounds can be significantly strengthened by use of cutting-plane methods. In doing so, we obtain a so-called "Branch-&-Cut" method.

**Example (Branch-&-Bound using linear-programming relaxation).**

$$
(IP) \qquad
\begin{aligned}
\max\ & -x_1 + x_2 \\
\text{subject to:} & \\
& 12x_1 + 11x_2 \le 63 \\
& -22x_1 + 4x_2 \le -33 \\
& x_1, x_2 \ge 0 \\
& x_1, x_2 \in \mathbf{Z}
\end{aligned}
$$

We solve the linear-programming relaxation of the initial subprogram and obtain an optimal solution with $z^* = 1.29$, $x_1^* = 2.12$, and $x_2^* = 3.41$.

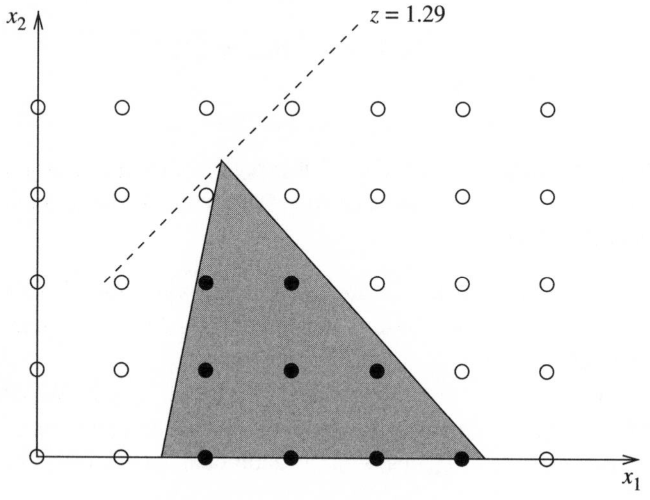

At this point we have

$$LB = -\infty, \ UB = 1.29,$$

and the list of active subprograms is

| Subprogram | $z^*$ | $x_1^*$ | $x_2^*$ |
|---|---|---|---|
| *IP* | 1.29 | 2.12 | 3.41 |

Selecting the only subprogram on the list, we arbitrarily choose $x_1$ as the variable to branch on. We obtain two new subprograms. The child with the constraint $x_1 \leq 2$ has $z^* = 0.75, x_1^* = 2.00, x_2^* = 2.75$ as the solution of its linear-programming relaxation. The child with the constraint $x_1 \geq 3$ has $z^* = -0.55, x_1^* = 3.00, x_2^* = 2.45$ as the solution of its linear-programming relaxation.

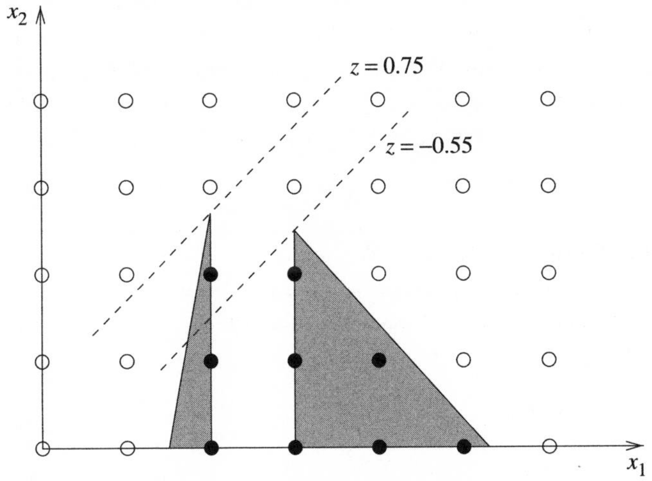


Both subprograms are put on the list. At this point we have

$$LB = -\infty, \ UB = 0.75,$$

and the list of active subprograms is

| Subprogram | $z^*$ | $x_1^*$ | $x_2^*$ |
|---|---|---|---|
| *IP* with $x_1 \leq 2$ | 0.75 | 2.00 | 2.75 |
| *IP* with $x_1 \geq 2$ | −0.55 | 3.00 | 2.45 |

Arbitrarily selecting the top subprogram from the list, we must branch on $x_2$, and we obtain two new subprograms. The child with the constraint $x_2 \leq 2$

has $z^* = 0.14$, $x_1^* = 1.86$, $x_2^* = 2.00$ as the solution of its linear-programming relaxation. This subprogram is placed on the list. The child with the constraint $x_2 \geq 3$ is fathomed by infeasibility. At this point we have:

$$LB = -\infty, \ UB = 0.14,$$

and the list of active subprograms is

| Subprogram | $z^*$ | $x_1^*$ | $x_2^*$ |
|---|---|---|---|
| *IP* with $x_1 \geq 3$ | −0.55 | 3.00 | 2.45 |
| *IP* with $x_1 \leq 2$, $x_2 \leq 2$ | 0.14 | 1.86 | 2.00 |

Arbitrarily selecting the bottom subprogram to remove from the list, we must branch on $x_1$, and we obtain two new subprograms. The child with the constraint $x_1 \leq 1$ is fathomed by infeasibility. The child with the constraint $x_1 \geq 2$ has $z^* = 0.00$, $x_1^* = 2.00$, $x_2^* = 2.00$ as the solution of its linear-programming relaxation and it is fathomed by optimality. This becomes our new candidate and we update *LB* to 0.00. Now, the remaining subprogram on the list (*IP* with $x_1 \geq 3$) is removed, and it is fathomed by bounds. At this point the list is empty and *UB* is reduced to 0.00, so we conclude that the candidate is optimal. ♠

**Exercise (Branch-&-Bound).** You are in the process of solving an integer linear maximization program, in the variables $x_1, x_2, x_3$, by Branch-&-Bound. The current value of the lower bound *LB* is $-\infty$. The list of active subproblems is

| Subprogram | $z^*$ | $x_1^*$ | $x_2^*$ | $x_3^*$ |
|---|---|---|---|---|
| *IP* with $x_1 \geq 6$, $x_2 \leq 3$ | 90.50 | 6.00 | 3.00 | 0.50 |
| *IP* with $x_1 \leq 5$, $x_2 \leq 13$ | 165.25 | 5.00 | 13.00 | 5.75 |
| *IP* with $x_1 \leq 5$, $x_2 \geq 14$, $x_3 \geq 1$ | 138.00 | 4.25 | 16.00 | 1.00 |
| *IP* with $x_1 \leq 5$, $x_2 \geq 14$, $x_3 \leq 0$ | 121.25 | 3.75 | 15.25 | 0.00 |

where $x^*$ is the optimal solution for the linear-programming relaxation of a subproblem and $z^*$ is the objective value of $x^*$.

a. What is the current value of the upper bound *UB*? Explain.
b. Have we fathomed any subproblem by integrality yet? Explain.
c. Have we fathomed any subproblem by bounds or by infeasibility yet? Explain.

**Exercise [Knapsack program, continued (see p. 82)].** Using the data from the "Knapsack program" Exercise, solve the knapsack program by using Branch-&-Bound.

There are some practical issues that can have a considerable effect on the performance of the Branch-&-Bound method for integer linear programming that has been described:

1. As is always true when the Branch-&-Bound framework is used, a greater value of *LB* increases the likelihood of fathoming by bounds. For integer linear programming, performance in this regard can be considerably enhanced by use of problem-specific heuristic methods to find a good initial candidate, rather than just hoping that one will be stumbled upon during the solution of the linear-programming relaxation of some subprogram.
2. All subprogram relaxations, except possibly for the initial one, may be solved by the dual simplex method, beginning with an optimal basis of the parent subprogram. This can drastically reduce the amount of time spent solving linear programs.

If the feasible region is not bounded, then it is possible that the linear-programming relaxation of *IP* may be unbounded. In such a case, however, *IP* is also unbounded if *IP* has a feasible solution and the data is rational (see the "Unbounded integer program" Problem from Chapter 6). Therefore, if the linear-programming relaxation of *IP* is unbounded, we determine whether *IP* is unbounded by applying Branch-&-Bound to the "feasibility program":

$$z_{IP_0} := \max x_0$$

subject to:

$$(IP_0) \qquad x_0 + \sum_{j=1}^{n} a_{ij} x_j \le b_i \ , \ \text{for } i = 1, 2, \ldots, m;$$

$$x_0 \le 0;$$

$$x_j \in \mathbf{Z} \ , \ \text{for } j = 0, 1, 2, \ldots, n.$$

*IP* is feasible if and only if $z_{IP_0} = 0$.

Branch-&-Bound can behave quite poorly on programs even when the feasible region is nicely bounded.

**Problem (Exponential example for Branch-&-Bound).** Let $n$ be an odd positive integer, and consider the integer linear program

$$\max \ -x_0$$

subject to:

$$x_0 + 2\sum_{j=1}^{n} x_j = n;$$

$$0 \le x_j \le 1, \text{ for } j = 0, 1, 2, \ldots, n;$$

$$x_j \in \mathbf{Z}, \text{ for } j = 0, 1, 2, \ldots, n.$$

Show that, when Branch-&-Bound is applied to this integer program, at least $2^{\frac{n-1}{2}}$ subprograms are placed on the list.

## 7.2 Knapsack Programs and Group Relaxation

The application of Branch-&-Bound to the solution of integer linear programs is not critically linked to the use of upper bounds obtained from the solution of linear-programming relaxations. For simplicity, consider the equality knapsack program

$$\max \sum_{j=1}^{n} c_j x_j$$

subject to:

$$(EKP) \qquad \sum_{j=1}^{n} a_j x_j = b;$$

$$x_j \ge 0, \text{ for } j = 1, 2, \ldots n;$$

$$x_j \in \mathbf{Z}, \text{ for } j = 1, 2, \ldots n,$$

where the $a_j$ and $b$ are arbitrary positive integers.

1. *Upper Bounds*: Choose $k$ ($1 \le k \le n$) so that $c_k/a_k = \max\{c_j/a_j \ : \ 1 \le j \le n\}$. Relaxing the *nonnegativity* restriction on the variable $x_k$, we obtain the *group relaxation*

$$\frac{c_k}{a_k} b + \max \sum_{j \ne k} \left( c_j - \frac{c_k}{a_k} a_j \right) x_j$$

subject to:

$$(GP) \qquad \sum_{j \ne k} a_j x_j = b - a_k x_k;$$

$$x_j \ge 0, \ \forall \ j \ne k;$$

$$x_j \in \mathbf{Z}, \text{ for } j = 1, 2, \ldots n.$$

Equivalently, we have

$$\frac{c_k}{a_k}b - \min \sum_{j \neq k} \left( -c_j + \frac{c_k}{a_k} a_j \right) x_j$$

(*GP*′)

subject to:

$$\sum_{j \neq k} a_j x_j \equiv b \ (\mathrm{mod}\ a_k);$$

$$x_j \geq 0, \ \forall\ j \neq k;$$

$$x_j \in \mathbf{Z}, \ \forall\ j \neq k.$$

Notice how, from every feasible solution to *GP*′, we can easily calculate the value for $x_k$ that goes with it in *GP*. If this value of $x_k$ is nonnegative, then we have found an optimal solution to *EKP*.

Now, consider a digraph $G$ with $V(G) := \{0, 1, \ldots, a_k - 1\}$. For each $i \in V(G)$ and $j \neq k$, we have an edge from $i$ to $i + a_j$ (mod $a_k$) with weight $-c_j + \frac{c_k}{a_k} a_j$. Notice that, by the choice of $k$, we have $-c_j + \frac{c_k}{a_k} a_j \geq 0$. Therefore, we have a digraph with nonnegative edge weights.

Consider a diwalk in $G$ that begins at vertex 0. As we traverse the walk, we increment variables starting out with $x_j = 0$, for $j = 1, 2, \ldots, n$. Including the edge from $i$ to $i + a_j$ (mod $a_k$) in our walk corresponds to incrementing the variable $x_j$ by one. Indeed, as $x_j$ is incremented by one, $\sum_{j \neq k} a_j x_j$ (mod $a_k$) changes from some value $i$ to $i + a_j$ (mod $a_k$), and $\sum_{j \neq k} \left( -c_j + \frac{c_k}{a_k} a_j \right) x_j$ increases by $-c_j + \frac{c_k}{a_k} a_j$. Ultimately, we want to choose a solution $x$ so that $\sum_{j \neq k} a_j x_j \equiv b$ (mod $a_k$), so we find a minimum-weight diwalk from vertex 0 to vertex $b$ (mod $a_k$). Because the edge weights are nonnegative, we can use Dijkstra's algorithm to find a minimum-weight di*path*.

**Example (Branch-&-Bound using group relaxation).** Consider the equality knapsack program (compare with p. 82):

$$\begin{array}{lcccccccccccl} \max & 11x_1 & + & 7x_2 & + & 5x_3 & + & x_4 & & & & & \\ \text{subject to:} & 6x_1 & + & 4x_2 & + & 3x_3 & + & x_4 & + & x_5 & = & 25; & \\ & x_1 & , & x_2 & , & x_3 & , & x_4 & , & x_5 & \geq & 0 & \text{integer.} \end{array}$$

We have $c_1/a_1 = \max\{c_j/a_j \ : \ 1 \leq j \leq 5\}$, so, relaxing the nonnegativity restriction on the variable $x_1$, we obtain the *group relaxation*

$$\begin{array}{lllcccccccccl} \frac{11}{6}25 & - & \min & \frac{1}{3}x_2 & + & \frac{1}{2}x_3 & + & \frac{5}{6}x_4 & + & \frac{11}{6}x_5 & & \\ & & \text{subject to:} & 4x_2 & + & 3x_3 & + & x_4 & + & x_5 & \equiv & 25 \ (\mathrm{mod}\ 6); \\ & & & x_2 & , & x_3 & , & x_4 & , & x_5 & \geq & 0 \ \text{integer.} \end{array}$$

The associated weighted digraph takes the following form:

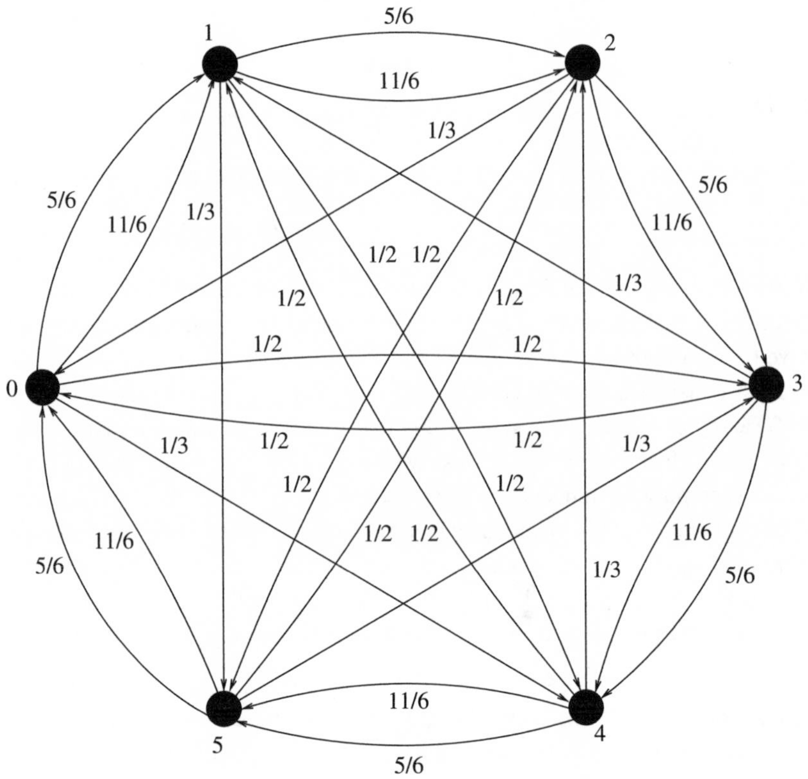


Because $1 \equiv 25 \pmod 6$, we seek a minimum-weight dipath from vertex 0 to vertex 1. The edge directly from vertex 0 to vertex 1 constitutes such a minimum-weight dipath. This corresponds to the solution $x_2 = 0$, $x_3 = 0$, $x_4 = 1$, $x_5 = 0$, and an upper bound of $45(= \frac{11}{6}25 - \frac{5}{6})$ for the group relaxation, which implies that $x_1 = 4$. Because $x_1$ turns out to be nonnegative, our solution to the group relaxation solves the original knapsack program.

Another minimum-weight dipath from vertex 0 to vertex 1 consists of the edge from vertex 0 to vertex 4 together with the edge from vertex 4 to vertex 1. This corresponds to the solution $x_2 = 1$, $x_3 = 1$, $x_4 = 0$, $x_5 = 0$, and the same upper bound of $45(= \frac{11}{6}25 - \frac{1}{3} - \frac{1}{2})$ for the group relaxation, which implies that $x_1 = 3$. Again, because $x_1$ turns out to be nonnegative, our solution to the group relaxation provides an alternative solution to the original knapsack program. ♠

**Problem (Big right-hand side).** Prove that if $b \geq (a_k - 1) \cdot \max\{a_j \,:\, j \neq k\}$, then the solution of the group relaxation solves the original knapsack program.

2. *Branching Rule*: Next, we need to see how we can branch effectively when we use the group relaxation. A subprogram will have the general form

$$
(EKP(l)) \qquad
\begin{aligned}
&\max \sum_{j=1}^{n} c_j x_j \\
&\text{subject to:} \\
&\sum_{j=1}^{n} a_j x_j = b; \\
&\qquad x_j \geq l_j, \ \text{for } j = 1, 2, \ldots n; \\
&\qquad x_j \in \mathbf{Z}, \ \text{for } j = 1, 2, \ldots n,
\end{aligned}
$$

where $l = (l_1, l_2, \ldots, l_n)$ is a vector of nonnegative integers. Substituting $x'_j := x_j - l_j$, we have the equivalent program

$$
(EKP'(l)) \qquad
\begin{aligned}
&\sum_{j=1}^{n} c_j l_j + \max \sum_{j=1}^{n} c_j x'_j \\
&\text{subject to:} \\
&\sum_{j=1}^{n} a_j x'_j = b - \sum_{j=1}^{n} a_j l_j; \\
&\qquad x'_j \geq 0, \ \text{for } j = 1, 2, \ldots n; \\
&\qquad x'_j \in \mathbf{Z}, \ \text{for } j = 1, 2, \ldots n,
\end{aligned}
$$

We can apply the group-relaxation method directly to *EKP*′(*l*). Finally, we branch by (1) considering $x = l$ as a potential replacement for the current candidate, and (2) replacing *EKP*(*l*) with the $n$ programs $EKP(l + \mathbf{e}^j)$, where $\mathbf{e}^j$ is the $j$th standard unit vector for $j = 1, 2, \ldots n$. This is a finite procedure because the assumption that the $a_j$ are positive bounds the $x_j$ from above.

**Exercise (Knapsack program using group relaxation).** Solve the equality knapsack program,

$$
\begin{array}{rccccccc}
\max & 16x_1 & + & 7x_2 & & & & \\
\text{subject to:} & 11x_1 & + & 5x_2 & + & x_3 & = & 18; \\
 & x_1 & , & x_2 & , & x_3 & \geq & 0 \quad \text{integer},
\end{array}
$$

by using the Branch-&-Bound scheme based on group relaxation.

## 7.3 Branch-&-Bound for Optimal-Weight Hamiltonian Tour

Let $G$ be a simple graph. We are interested in the problem of finding a maximum-weight Hamiltonian tour of $G$.

1. *Upper Bounds*: We fix a vertex $w$ in $V(G)$. A *$w$-forest* of $G$ is a set that consists of no more than two edges from $\delta_G(w)$ and a forest of $G[V(G) - w]$. Similarly, a *$w$-tree* of $G$ is a set that consists of two edges from $\delta_G(w)$ and a spanning tree of $G[V(G) - w]$. It is easy to see that every Hamiltonian tour of $G$ is a $w$-tree of $G$, but not conversely. On the other hand, the only reason that a given $w$-tree is not a Hamiltonian tour is that some of the vertices, other than $w$, have a degree different from two.

Let $\mathcal{F}_w(G)$ [respectively, $\mathcal{T}_w(G)$] be the set of all $w$-forests (respectively, $w$-trees) of $G$. Let $\mathcal{P}_w(G)$ be the convex hull of the characteristic vectors of elements of $\mathcal{T}_w(G)$. We can formulate the problem of finding a maximum-weight Hamiltonian tour as

$$z := \max \sum_{e\in E(G)} c(e)x_e$$

subject to:

$$\sum_{e\in\delta_G(v)} x_e = 2\,, \quad \forall\, v \in V(G) - w \ (\textit{degree constraints});$$

$$x \in \mathcal{P}_w(G)\,.$$

It is easy to see that the set $\mathcal{F}_w(G)$ is the set of independent sets of a matroid $\mathcal{M}_w(G)$. Assuming that $G$ has a $w$-tree, the matroid $\mathcal{M}_w(G)$ is the direct sum of the uniform matroid of rank two on $\delta_G(w)$ and the graphic matroid of $G[V(G) - w]$. Furthermore, again assuming that $G$ has a $w$-tree, the set of bases of $\mathcal{M}_w(G)$ is the set $\mathcal{T}_w(G)$.

By dropping the degree constraints, we are led to

$$f := \max \sum_{e\in E(G)} c(e)x_e$$

subject to:

$$x \in \mathcal{P}_w(G)\,,$$

which is an upper bound on $z$. We can efficiently calculate $f$ by using the Greedy Algorithm to find a maximum-weight base of $\mathcal{M}_w(G)$.

**Example ($w$-tree relaxation).** We consider the maximum-weight Hamiltonian-tour problem on the graph of the "Maximum-weight spanning tree" Exercise (see p. 58). We take $w := b$. The maximum-weight $b$-tree is

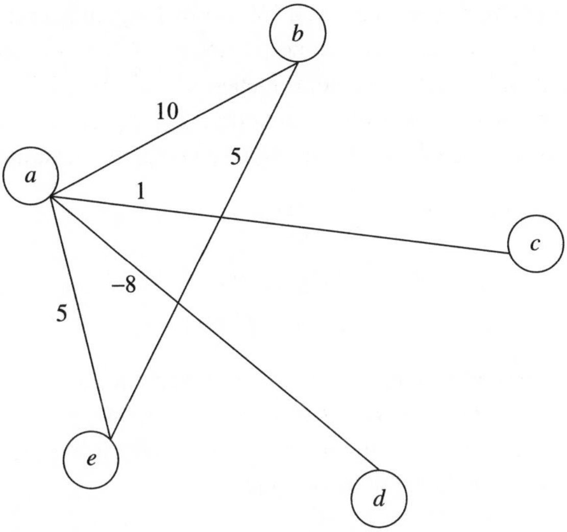


with weight $f = 13$. ♠

To use this bound $f$ in a Branch-&-Bound Algorithm, we need to choose a branching strategy that is consistent with the bound calculation.

2. *Branching Rule*: Every $w$-tree contains a unique cycle. If $S$ is a maximum-weight $w$-tree and $S$ is not a Hamiltonian tour, then $S$ contains a cycle $C$ with $|C| < |V(G)|$. Every Hamiltonian tour omits some edge of $C$; so we can replace the Hamiltonian-tour problem with the $|C|$ Hamiltonian-tour subproblems, which are form by taking the restriction of the problem to $G.(E(G) - e)$ for each $e \in C$.

**Example ($w$-tree relaxation, continued).** Continuing with the example, the maximum-weight $b$-tree contains the unique cycle $C := \{\{a, b\}, \{b, e\}, \{a, e\}\}$. It is easy to check that the maximum-weight $b$-tree on the graph with the edge $\{a, b\}$ deleted has weight 5, weight 10 on the graph with $\{b, e\}$ deleted, and weight 8 on the graph with $\{a, e\}$ deleted. The maximum of these is 10, so at this point in the Branch-&-Bound Algorithm, we have reduced the global upper bound from 13 to 10. ♠

**Exercise ($w$-tree based Branch-&-Bound).** Continuing with the "$w$-tree relaxation" Example, find the maximum-weight Hamiltonian tour by completing the execution of the Branch-&-Bound Algorithm.

The $w$-tree bound can be improved by use of Lagrangian relaxation (see p. 35). In this way, we can take some advantage of the heretofore ignored degree constraints. For notational convenience, we take $\pi \in \mathbf{R}^{V(G)}$, but we fix $\pi_w := 0$, and we work only with $\pi$ satisfying $\sum_{v \in V(G)} \pi_v = 0$ [initially, we take $\pi_v = 0$, for all $v \in V(G)$]. We have the Lagrangian relaxation

$$(L(\pi)) \qquad \begin{aligned} f(\pi) := \max & \sum_{\{i,j\} \in E(G)} [c(\{i,j\}) - \pi_i - \pi_j] x_{\{i,j\}} \\ & \text{subject to:} \\ & x \in \mathcal{P}_w(G) , \end{aligned}$$

Let $\widetilde{S}$ be a maximum-weight $w$-tree, with respect to the weight function: $c(\{i,j\}) - \pi_i - \pi_j$, for all $\{i,j\} \in E(G)$. We have the subgradient $\widetilde{h} \in \mathbf{R}^{V(G)}$, defined by $h_v = 2 - |S \cap \delta_G(v)|$, for all $v \in V(G)$. We can use the Subgradient Optimization Algorithm to seek to minimize $f$.

**Example ($w$-tree relaxation, continued).** We continue with the "$w$-tree relaxation" Example. At $\pi = \widetilde{\pi}^1 := 0$, the optimal $w$-tree is $\widetilde{S} = \{\{a,b\}, \{b,e\}, \{a,e\}, \{a,c\}, \{a,d\}\}$. We have the subgradient

$$\widetilde{h} = \begin{pmatrix} \widetilde{h}_a \\ \widetilde{h}_c \\ \widetilde{h}_d \\ \widetilde{h}_e \end{pmatrix} = \begin{pmatrix} -2 \\ 0 \\ 1 \\ 1 \end{pmatrix} .$$

This leads to the new iterate $\widetilde{\pi}^1 = \widetilde{\pi}^0 - \lambda \widetilde{h}$. The Lagrangian relaxation $L(\widetilde{\pi}^0 - \lambda \widetilde{h})$ is the problem of finding a maximum-weight $b$-tree on the edge-weighted graph:

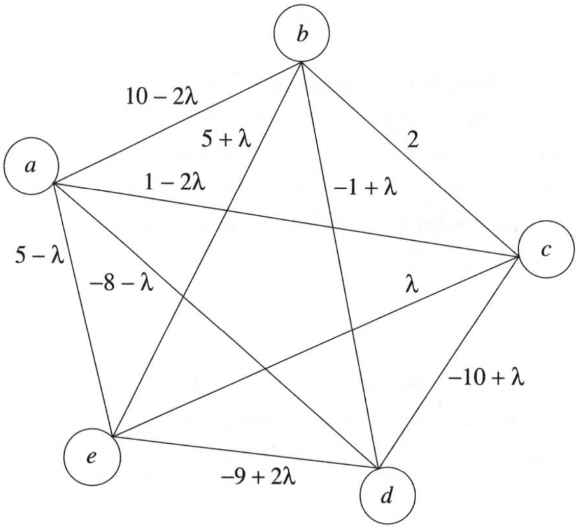

Plotting $f(\widetilde{\pi}^1) = f(\widetilde{\pi}^0 - \lambda\widetilde{h})$ as a function of $\lambda$, we have

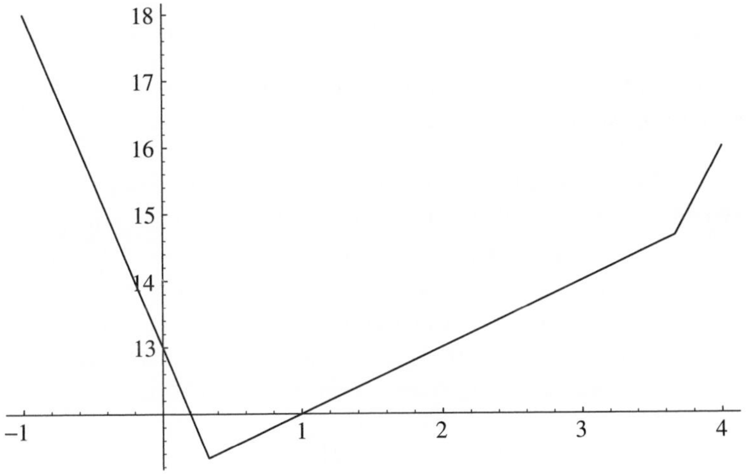


By taking $\lambda = 2/5$, we see that we can get the upper bound down from $f = 13$ (at $\lambda = 0$) to less than 12. Then, using the integrality of the weight function, we can conclude, without yet resorting to branching, that $z \leq 11$. ♠

There are other reasonable choices of bounds for carrying out Branch-&-Bound to find a maximum-weight Hamiltonian tour. For example, every Hamiltonian tour is a 2-factor, so the weight of a maximum-weight 2-factor is an upper bound on the weight of all Hamiltonian tours. We can find a maximum-weight 2-factor by solving a related maximum-weight perfect-matching problem (see p. 135). For a compatible branching rule, we proceed exactly as we did previously, branching on a cycle.

## 7.4 Maximum-Entropy Sampling and Branch-&-Bound

The Branch-&-Bound framework may also be applied to situations in which upper bounds are not easily calculated by solving a relaxation. One such example is the "maximum-entropy sampling problem." Let $C$ be a symmetric positive-definite matrix, with rows and columns indexed by $V$. For nonempty $S, T \subset V$, let $C[S, T]$ denote the submatrix of $C$ having rows indexed by $S$ and columns indexed by $T$. For nonempty $S \subset V$, the *entropy* of $S$ with respect to $C[V, V]$ is $H(S) := \ln \det(C[S, S])$. If $V$ is a set of random variables having a joint Gaussian distribution and $C$ is the associated covariance matrix, then $H(S)$ is a measure of the "information" contained in $S$.

Let $s \leq |V|$ be a positive integer, and let $\alpha$ be a constant. The *maximum-entropy sampling problem* is

$$P(C, V, s, \alpha) \qquad\qquad z := \alpha + \max\{H(S) \ : \ S \subset V, \ |S| = s\}.$$

The goal of the problem is to find a choice of $s$ of the random variables that is most informative about the entire set $V$.

1. *Upper Bounds*: A consequence of the "eigenvalue interlacing property" for symmetric matrices is that $\det(C[S, S]) \leq \prod_{l=1}^{s} \lambda_l(C[V, V])$, for all $S \subset V$ having $|S| = s$, where $\lambda_l()$ denotes the $l$th greatest eigenvalue. Therefore, we have the upper bound $z \leq \alpha + \sum_{l=1}^{s} \ln \lambda_l(C[V, V])$.
2. *Branching Rule*: We branch by choosing some $j \in V$ and creating two subproblems of the problem $P(C, V, s, \alpha)$. For one of them, we exclude $j$ from the optimal solution. This amounts to solving the problem

$$P(C[V - j, V - j], V - j, s, \alpha).$$

   For the other, we require $j$ to be in the optimal solution. This amounts to solving the problem

$$\begin{aligned} P(&C[V - j, V - j] - C[V - j, j]C_{jj}^{-1}C[j, V - j], \\ &V - j, s - 1, \alpha + \ln C_{jj}). \end{aligned}$$

3. *Lower Bounds*: We can obtain a lower bound on $z$ by some greedy and local-search heuristics.

**Problem (Maximum-entropy sampling and Branch-&-Bound).** Let $V_1, V_2, \ldots, V_p$ be a partition of $V$. Let $C(V_1, V_2, \ldots, V_p)$ be the submatrix of $C$ we obtain by changing $C_{ij}$ to zero whenever $i$ and $j$ are in different parts of $V_1, V_2, \ldots, V_p$.

a. Prove that

$$z \leq \alpha + \sum_{l=1}^{s} \ln \lambda_l \left(C(V_1, V_2, \ldots, V_p)\right).$$

b. Prove that $z$ is equal to the optimal objective value of

$$P(C^{-1}, V, n - s, \alpha + \ln\det(C)),$$

and explain how this formula can be exploited, in conjunction with the upper bounds from part a, to give further upper bounds on $z$.

## 7.5 Further Study

There are many papers on Branch-&-Bound. In most of these, the mathematics is mostly in the upper-bounding methods (for maximization problems). Lenstra (1983) and Margot (2003) demonstrate (in quite different ways) the value of mathematical approaches to the investigation of branching rules in the context of integer linear programming.

Branch-&-Bound for integer linear programming has several, more sophisticated variants. Some of these are Branch-&-Cut, Branch-&-Price and Branch-Cut-&-Price. See Ladányi, Ralphs, and Trotter (2001) for a description of many of the ideas and how they have been implemented. Notably, open-source code for Branch-Cut-&-Price is available at www.coin-or.org.

There is considerable practical lore on applying Branch-&-Bound (and its variants) to integer linear programming problems. The articles by Linderoth and Savelsbergh (1999) and Martin (2001) are in this vein.

An entry point to the literature on Branch-&-Bound methods for maximum-entropy sampling is Lee (2001).

Anstreicher, Brixius, Goux, and Linderoth (2002) achieved stunning success in solving large "quadratic assignment problems" by employing sophisticated bounding procedures, clever branching rules, and a state-of-the-art parallel implementation on a computational grid.

# 8

## *Optimizing Submodular Functions*

Minimizing and maximizing submodular functions are fundamental unifying problems in combinatorial optimization. In this chapter, some examples are given, and we discuss aspects of the general problems of minimizing and maximizing submodular functions.

### 8.1 Minimizing Submodular Functions

Let $M$ be a matroid. Recall the rank inequalities

$$\sum_{e \in S} x_e \le r_M(S), \quad \forall\, S \subset E(M)$$

that, along with nonnegativity, describe $\mathcal{P}_{\mathcal{I}(M)}$. The separation problem for the rank inequalities is, for fixed $x^* \in \mathbf{R}^{E(M)}$, find $S \subset E(M)$ so that

$$\sum_{e \in S} x^*_e > r_M(S) .$$

Define $f : 2^{E(M)} \mapsto \mathbf{R}$ by

$$f(S) := r_M(S) - \sum_{e \in S} x^*_e .$$

It is easy to check that $f$ is submodular (by use of the fact that $r_M$ is). Moreover, $x^*$ violates a rank inequality if and only if the minimum of $f(S)$, over $S \subset E(M)$, is less than 0.

Thus an ability to minimize this particular submodular function efficiently, provides a *theoretically* efficient algorithm, by use of the ellipsoid method, for finding maximum-weight sets that are independent for a matroid or even for a pair of matroids. Of course, we also know direct combinatorial algorithms for these problems that are practically as well as theoretically efficient.

**Problem (Minimum-weight cuts and submodular minimization).** Consider a digraph $G$ with distinguished vertices $v, w \in V(G)$ with $v \neq w$, and a "capacity" function $c : E(G) \to \mathbf{R}_+$. For $S \subset V(G) \setminus \{v, w\}$, define

$$f(S) := \sum \left\{c(e) \ : \ e \in \delta^+_G(S + v)\right\}.$$

[That is, $f(S)$ is the sum of the capacities on the edges that point *out* of $S + v$.] Prove that $f$ is submodular, and describe how to determine the minimum of $f$ on $V(G) \setminus \{v, w\}$.

**Problem (Maximum-cardinality matroid intersection and submodular minimization).** Let $M_i$ be matroids on the common ground set $E := E(M_i)$ for $i = 1, 2$. Prove that $f : 2^E \mapsto \mathbf{R}$ defined by

$$f(S) := r_{M_1}(S) + r_{M_2}(E \setminus S)$$

is submodular, and explain how this relates to the problem of finding a maximum-cardinality subset of $E$ that is independent in both $M_1$ and $M_2$.

Next, we discuss some aspects of the problem of *minimizing* a *general* submodular function $f : 2^E \mapsto \mathbf{R}$, where $E$ is a finite set. First, we may assume that $f(\emptyset) = 0$ [by subtracting $f(\emptyset)$ from $f$ if necessary]. We define $f' : [0, 1]^E \mapsto \mathbf{R}$ in a certain way, so that $f'(x) := f(S(x))$ for $x \in \{0, 1\}^E$. Every nonzero $x \in [0, 1]^E$ can be decomposed uniquely as $x = \sum_{j=1}^m \lambda_j x^j$, where

| | |
|---|---|
| *(i)* | $m \leq \|E\|;$ |
| *(ii)* | $\lambda_j > 0, \text{for} j = 1, 2, \ldots m;$ |
| *(iii)* | $x^j \in \{0, 1\}^E, \text{for } j = 1, 2, \ldots m;$ |
| *(iv)* | $x^1 \geq x^2 \geq \cdots \geq x^m \neq 0.$ |

Then we let $f'(x) := \sum_{j=1}^m \lambda_j f(S(x^j))$.

**Theorem (Convexity of $f'$ and integer-valued minima).** *The function $f'$ is convex and attains it minimum over* $[0, 1]^E$ *on* $\{0, 1\}^E$.

*Proof.* First, we demonstrate that the function $f'$ is convex. Consider a point $x^* \in \mathbf{R}_+^E$ and the linear program

$$\hat{f}(x^*) := \max \sum_{e \in E} x_e^* z_e$$

Subject to:

$$\sum_{e \in T} z_e \leq f(T), \quad \forall\, T \subset E\,.$$

The optimal objective-function value of a linear-programming maximization problem is a convex function of the vector of objective-function coefficients. Therefore, it suffices to prove that $f' = \hat{f}$.

Without loss of generality, we can take $E := \{1, 2, \ldots, n\}$ and $x_1^* \geq x_2^* \geq \cdots \geq x_n^*$. Let $T_j := \{1, 2, \ldots, j\}$, for $j = 1, 2, \ldots, n$, and let $T_0 := \emptyset$. The proof of the characterization of $\mathcal{P}_{\mathcal{I}(M)}$ for a matroid $M$ implies that

$$\begin{aligned}\hat{f}(x^*) &= \sum_{j=1}^{n} x_j^* \Big[ f(T_j) - f(T_{j-1}) \Big] \\ &= \sum_{j=1}^{n} \Big( x_j^* - x_{j+1}^* \Big) f(T_j)\end{aligned}$$

(even though $f$ need not be the rank function of a matroid), where we take $x_{n+1}^* := 0$. Letting $\lambda_j := x_j^* - x_{j+1}^*$, we get the decomposition $x^* = \sum_{j=1}^{n} \lambda_j x(T_j)$ (we can ignore the $j$ with $\lambda_j = 0$); so we have $f'(x^*) = \hat{f}(x^*)$.

Finally, we demonstrate that $f'$ is minimized at a vertex of $[0, 1]^E$. Let $x^* = \sum_{j=1}^{m} \lambda_j x^j \in [0, 1]^E$ be a minimizer of $f'$ over $[0, 1]^E$. If $f'(x^*) = 0$, then $f'$ is also minimized by $0 \in \{0, 1\}^E$, because we have assumed that $f(\emptyset) = f'(0) = 0$. Therefore, we may suppose that $f'(x^*) < 0$. If $f'(x^j) > f'(x^*)$ for all $j$, then $f'(x^*) = \sum_{j=1}^{m} \lambda_j f'(x^j) > \sum_{j=1}^{m} \lambda_j f'(x^*)$. Because $f'(x^*) < 0$, we have $1 < \sum_{j=1}^{m} \lambda_j$. However, $x^* \in [0, 1]^E$ implies that $\sum_{j=1}^{m} \lambda_j \leq 1$, so we have a contradiction. Therefore, $f'(x^j) \leq f'(x^*)$ for some $j$; hence, some $x^j$ minimizes $f'$. ∎

There is a theoretically efficient algorithm for minimizing the convex function $f'$ over $[0, 1]^E$, by use of the ellipsoid method. In this way, we can find a minimum of the submodular function $f$. Other, more combinatorial methods generalize maximum-flow techniques.

## 8.2 Minimizing Submodular Functions Over Odd Sets

In this section, we see how to use an efficient algorithm for minimizing a submodular function as a subroutine in an efficient algorithm for minimizing a submodular function $f$ over subsets $S$ of the ground set $E$ intersecting a fixed subset $T$ of the ground set on an odd number of elements. First, some motivation is given that is related to the maximum-weight matching problem.

Let $H$ be a graph with weight function $c$ on $E(H)$. We consider the maximum-weight matching problem on $H$. We may as well assume that $c$ is nonnegative, as the set of matchings is an independence system and so no negative-weight edge will appear in a maximum-weight matching. Next, we can make a copy $H'$ of $H$, and join each $i \in V(H)$ to its copy $i' \in V(H')$. Call this new graph $G$. All edges of $H'$ and those extending between $H$ and $H'$ are assigned 0 weight as we extend $c$ to the entire edge set of $G$.

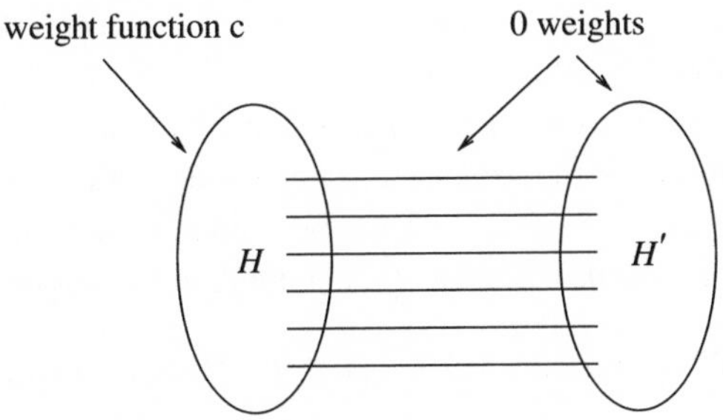


Now, every matching $S$ of $H$ extends to a perfect matching of $G$ having the same weight – for each $e \in S$, take its copy $e' \in E(H')$, and for each exposed vertex $i$ of $H$, take the edge joining $i$ to its copy $i' \in H'$. Furthermore, every perfect matching $S$ of $G$ induces a matching $S \cap E(H)$ of $H$ having the same weight. Therefore, to efficiently find a maximum-weight matching of $H$, it suffices to be able to find a maximum-weight perfect matching of $G$.

Therefore, let us assume now that we have an arbitrary graph $G$ and nonnegative-weight function $c$ on $E(G)$. Let $\overline{\mathcal{M}}(G)$ be the set of perfect matchings of $G$. Considering the inequality characterization of the matching polytope $\mathcal{P}_{\mathcal{M}(G)}$ (see the Matching Polytope Theorem, p. 109), it is easy to see that the perfect-matching polytope $\mathcal{P}_{\overline{\mathcal{M}}(G)}$ is the solution set of

$$(i) \qquad -x_e \le 0, \quad \forall\, e \in E(G);$$

$$(ii) \qquad \sum_{e \in \delta_G(v)} x_e = 1, \quad \forall\, v \in V(G);$$

$$(iii) \qquad \sum_{e \in E(G[W])} x_e \le \frac{|W|-1}{2}, \quad \forall\, W \subset V(G) \text{ with } |W| \ge 3 \text{ odd}.$$

Using equation *(ii)*, it is easy to check that *(iii)* can be replaced with

$$(iii') \qquad \sum_{e \in \delta_G(W)} x_e \geq 1, \quad \forall\, W \subset V(G) \text{ with } |W| \text{ odd};$$

we simply note that

$$2\left(\sum_{e\in E(G[W])} x_e \leq \frac{|W|-1}{2}\right) - \sum_{v\in W}\left(\sum_{e\in\delta_G(v)} x_e = 1\right) = -\left(\sum_{e\in\delta_G(W)} x_e \geq 1,\right).$$

Then, for $x^* \in \mathbf{R}^{E(G)}$ satisfying $(i)$, we can solve the separation problem for $(iii')$ by minimizing the function

$$f(W) := -1 + \sum_{e\in\delta_G(W)} x_e^*$$

over odd cardinality subsets $W$ of $V(G)$. As we have seen ("Cuts" Problem, p. 61), the function $f$ is submodular.

Now, we turn back to the general problem of minimizing a submodular function over subsets of the ground set intersecting a fixed subset on an odd number of elements. To be precise, let $f$ be a submodular function on $E$. Let $T$ be a subset of $E$. We describe an efficient method for solving

$$X^* := \operatorname{argmin}\{f(X) \ : \ X \subset E,\ |X \cap T| \text{ odd}\}.$$

We assume that we have, at our disposal, an efficient subroutine for ordinary submodular-function minimization.

*Step 1: Reduce the case of $|T|$ odd to $|T|$ even.* We observe that an optimal $X^*$ either contains all of $T$ or it avoids some element of $T$. Therefore, we calculate

$$X_T := \operatorname{argmin}\{f(X) \ : \ X \subset E,\ T \subset X\},$$

and, for all $e \in T$,

$$X_e := \operatorname{argmin}\{f(X) \ : \ X \subset E - e,\ |X \cap (T - e)| \text{ odd}\}.$$

The calculation of each $X_e$ is just like the calculation of $X^*$, but now we are intersecting an even cardinality set on an odd number of elements. The calculation of $X_T$ is just an instance of ordinary submodular-function minimization, but the effective ground set is just $E \setminus T$, as we can just "shrink" $T$ to a single element.

*Step 2: Solve a relaxation.* Let $U$ be a subset of $E$. We say that $U$ *splits* $T$ if both $T \cap U$ and $T \setminus U$ are nonempty. We wish to calculate

$$U := \operatorname{argmin}\{f(X) \ : \ X \subset E, \ X \text{ splits } T\}.$$

Notice that, because $T$ is even, the condition that $X$ splits $T$ is a weakening of the condition that $|X \cap T|$ is odd. Therefore, if $|U \cap T|$ is odd, then we solve our original problem by letting $X^* = U$.

Next, we specify how we can efficiently calculate $U$. For all distinct $e, f \in T$, we calculate

$$U_{e,f} := \operatorname{argmin}\{f(X) \ : \ X \subset E - f, \ e \in X \text{ odd}\}.$$

The calculation of each of the $\binom{|T|}{2}$ choices of $U_{e,f}$ is just an ordinary submodular-function minimization problem. Then we simply let

$$U = \operatorname{argmin}\left\{f(U_{e,f}) \ : \ e, f \in T\right\}.$$

*Step 3: Recurse.* At this point we can assume that $|U \cap T|$ (and hence, also $|T \setminus U|$) is even, or we would have solved the problem in Step 2. Recursively, we solve the following two subproblems:

$$U_1 := \operatorname{argmin}\{f(X) : X \subset E, \ |X \cap (T \cap U)| \text{odd}, \ \ X \text{ does not split } T \setminus U\};$$
$$U_2 := \operatorname{argmin}\{f(X) : X \subset E, \ |X \cap (T \setminus U)| \text{ odd}, \ \ X \text{ does not split } T \cap U\}.$$

Although we still have some work left to do to justify this, the solution to our main problem is just to set $X^*$ to

$$\operatorname{argmin}\{f(U_1), f(U_2)\}.$$

We note that, for the calculations of $U_1$ and $U_2$, we reduce the problems to problems of the same type that we are attacking, by "shrinking" the set not to be split to a single element. In doing so, the set that we need to intersect on an odd cardinality set remains of even cardinality. Also, because $|T| = |T \setminus U| + |T \cap U|$, it is clear that the total number of recursive calls will be less than $T$.

**Theorem (Correctness for odd submodular minimization).** *If $|U \cap T|$ is even, then $X^* = \operatorname{argmin}\{f(U_1), f(U_2)\}$ solves*

$$\min\{f(X) \ : \ X \subset E, \ |X \cap T| \ \textit{odd}\}.$$

*Proof.* The proof is by contradiction. Suppose that $X^*$ solves our main problem, but $f(X^*) < f(U_1)$ and $f(X^*) < f(U_2)$. From the definition of $U_1$, we see that $X^*$ splits $T \setminus U$. Therefore, immediately, we have that $X^* \cup U$ splits $T$. Symmetrically, from the definition of $U_2$, we see that $X^*$ splits $T \cap U$. Therefore, we have that $X^* \cap U$ splits $T$.

Now, because $|T \cap X^*|$ is odd and $|T \cap U|$ is even, exactly one of $|T \cap (X^* \cup U)|$ and $|T \cap (X^* \cap U)|$ is odd. We suppose that $|T \cap (X^* \cap U)|$ is odd (the other case, which is left for the reader, is handled symmetrically). By the definition of $X^*$, we have

$$f(X^* \cap U) \geq f(X^*).$$

By the definition of $U$, we have

$$f(X^* \cup U) \geq f(U).$$

Then, by submodularity, we must have $f(X^* \cap U) = f(X^*)$ [and $f(X^* \cup U) = f(U)$]. Now, $(X^* \cap U) + (T \cap U) = (X^* \cap U) \cap T$. Therefore, $|(X^* \cap U) + (T \cap U)|$ is odd and $(X^* \cap U) + (T \cap U)$ does not split $T \setminus U$. Therefore, by the definition of $U_1$, we have $f(X^* \cap U) \geq f(U_1)$. However, because we have already established that $f(X^* \cap U) = f(X^*)$, we conclude that $f(X^*) \geq f(U_1)$, which contradicts our assumption. ∎

## 8.3 Maximizing Submodular Functions

For an interesting special case, we know an efficient algorithm for maximizing a submodular function.

**Problem (Maximum-cardinality matching and submodular maximization).** Let $G$ be an undirected graph and define $f : 2^{E(G)} \mapsto \mathbf{R}$ by

$$f(S) := |\{v \in V(G) \ : \ e \in \delta_G(v) \text{ for some } e \in S\}| - |S| ,$$

for $S \subset E(G)$. Prove that $f$ is submodular and that, if $S$ maximizes $f$, then $f(S)$ is the maximum number of edges in a matching of $G$.

In general, maximizing a submodular function is hard. For example, the difficult problem of determining whether a digraph has a directed Hamiltonian tour is a problem of finding a maximum-cardinality set that is independent for *three* matroids having a common ground set.

**Problem (Maximum-cardinality $p$-matroid intersection and submodular maximization).** Let $M_i$ be matroids on the common ground set $E :=$

$E(M_i)$, for $i = 1, 2, \ldots, p$. Define a submodular function $f : 2^E \mapsto \mathbf{R}$ by

$$f(S) := \sum_{i=1}^{p} r_{M_i^*}(S) .$$

Prove that the problem of finding a maximum-weight set that is independent in all $p$ matroids can be recast as a problem of finding a set $S$ that maximizes $f(S)$.

Hard submodular maximization problems arise in other domains as well.

**Problem (Uncapacitated facility location and submodular maximization).** Recall the "Uncapacitated facility location" Problem (see p. 6). Demonstrate how the uncapacitated facility-location problem can be modeled as a problem of maximizing a submodular function.

Another favorite hard problem can also be modeled as a problem of maximizing a submodular function.

**Problem (Vertex packing and submodular maximization).**

a. Recall the entropy function $H$ (see p. 191). Prove that $H$ is a submodular function.
b. Let $G$ be a simple graph. Let $C$ be the symmetric matrix, with row and columns indexed by $V(G)$, having

$$c_{ij} := \begin{cases} 1, & \text{if } \{i, j\} \in E(G) \\ 0, & \text{if } i \neq j \text{ and } \{i, j\} \notin E(G). \\ 3|V(G)|, & \text{if } i = j \end{cases}$$

The matrix $C$ is symmetric and positive definite. Notice that if $E(G[S]) = \emptyset$, then $H(S) = |S| \cdot \ln(3|V(G)|)$. Prove that if $E(G[S]) \neq \emptyset$, then $H(S) < |S| \cdot \ln(3|V(G)|)$.

## 8.4 Further Study

The first theoretically efficient algorithm for minimizing a submodular function was based on the ellipsoid method; see Grötschel, Lovász and Schrijver (1988). The first theoretically efficient combinatorial algorithms are due (simultaneously!) to Iwata, Fleischer, and Fujishige (1999) and to Schrijver (2000).

None of these algorithms should be regarded as practical. However, their existence, together with the known practical and theoretically efficient algorithms for the minimization of particular submodular functions, suggests that it is useful to know whether a particular combinatorial-optimization problem can be regarded as a problem of minimizing a submodular function.

The work by McCormick (2003) is a very nice survey of the state-of-the-art algorithms for minimizing submodular functions.

> "It is true what Madame says," observed Jacques Three. "Why stop? There is great force in that. Why stop?"
>
> "Well, well," reasoned Defarge, "but one must stop somewhere. After all, the question is still where?"
>
> – *A Tale of Two Cities (C. Dickens)*

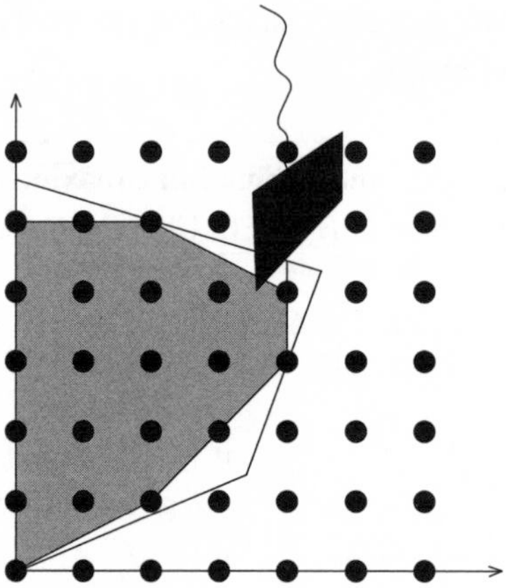

# *Appendix: Notation and Terminology*

Some familiarity with set operations, matrices and vector spaces, graphs, and digraphs is assumed. In this appendix, basic notation and terminology that we make free use of are given

## A.1 Sets

(In what follows, $S$, $T$ are subsets of the finite ground set $E$.)

| Notation / Term | Definition |
|---|---|
| $e \in S$ (*in*) | $e$ is an element of $S$ |
| $S \subset T$ (*subset*) | $e \in S \Rightarrow e \in T$ |
| $S \cap T$ (*intersect*) | $\{e : e \in S \text{ and } e \in T\}$ |
| $S \cup T$ (*union*) | $\{e : e \in S \text{ or } e \in T\}$ |
| $S + e$ (*plus*) | $S \cup \{e\}$ (assumes $e \notin S$) |
| $S \setminus T$ (*minus*) | $\{e \in S : e \notin T\}$ |
| $S - e$ (*minus*) | $S \setminus \{e\}$ (assumes $e \in S$) |
| $S \Delta T$ (*symmetric difference*) | $(S \setminus T) \cup (T \setminus S)$ |
| $\|S\|$ (*cardinality*) | number of elements in $S$ |
| $2^S$ (*power set*) | $\{X : X \subset S\}$ |

## A.2 Algebra

(In what follows, $A$ is a matrix with elements from field $\mathbf{F}$ with $m$ rows and columns indexed from finite set $E$.)

| Notation / Term | Definition |
|---|---|
| $\mathbf{R}$ | the reals |
| $\mathbf{R}_+$ | the nonnegative reals |

| Notation / Term | Definition |
|---|---|
| **Q** | the rationals |
| **GF(p)** | the Galois field with $p$ elements ($p$ a prime power) |
| **Z** | the integers |
| $\mathbf{Z}_+$ | the nonnegative integers |
| **e** | vector/matrix with all components equal to 1 |
| $\mathbf{e}^i$ | $i$th standard unit vector |
| $I$ | identity matrix |
| $A_S$ | matrix consisting of columns of $A$ indexed by $S \subset E$ |
| rank($A$) | number of linearly independent rows (columns) of $A$ |
| det($A$) | determinant of $A$ (assumes $A$ square) |
| $A^T$ | transpose of $A$ |
| $\mathbf{F}^E$ | set of points in $\mathbf{F}^{\lvert E \rvert}$ with coordinates indexed from $E$ |
| r.s.($A$) (*row space*) | $\{x \in \mathbf{F}^E : x^T = y^T A,\ y \in \mathbf{F}^m\}$ |
| c.s.($A$) (*column space*) | r.s.($A^T$) |
| n.s.($A$) (*null space*) | $\{x \in \mathbf{F}^E : Ax = 0\}$ |
| $x(S)$ | characteristic vector of $S \subset E$ |
| $S(x)$ (*support*) | $\{e \in E : x_e \neq 0\}$ (where $x \in \mathbf{F}^E$) |
| $\lVert x \rVert$ (*2-norm*) | $\sqrt{\sum_{j \in E} x_j^2}$ (where $x \in \mathbf{R}^E$) |

## A.3 Graphs

(In what follows, $G$ is a finite undirected graph.)

| Notation / Term | Definition |
|---|---|
| $V(G)$ | the vertex set of $G$ |
| $E(G)$ | the edge set of $G$ |
| $\kappa(G)$ | number of connected components of $G$ (counting isolated vertices) |
| $\delta_G(v)$ | the edges having $v$ as exactly one endpoint [$v \in V(G)$] |
| $\delta_G(S)$ | the edges having exactly one endpoint in $S$ [$S \subset V(G)$] |
| $N(S)$ (*neighbors of* $S$) | vertices in $V(G) \setminus S$ adjacent to some vertex in $S$ [$S \subset V(G)$] |
| $G[S]$ (*induced subgraph*) | $V(G[S]) := S$, $E(G[S]) :=$ edges in $E(G)$ with both ends in $S \subset V(G)$ |

| Notation / Term | Definition |
|---|---|
| $G.F$ (*restriction*) | $V(G.F) := V(G)$, $E(G.F) := F$ $[F \subset E(G)]$ |
| *cycle* | simple closed path of edges |
| *forest* | a set of edges containing no cycle |
| *spanning tree* | a forest $F$ (of $G$) such that $\kappa(G.F) = 1$ |
| *loop* | cycle of one edge |
| *simple graph* | no loops or identical edges |
| *cocycle* | (set-wise) minimal disconnecting edge-set |
| *coloop* | a cocycle of one edge |
| $A(G)$ | 0/1-valued vertex-edge incidence matrix of $G$ |
| $K_n$ | complete graph on $n$ vertices |
| $K_{m,n}$ | complete bipartite graph with parts of $m$ and $n$ vertices |
| *matching* | $F \subset E(G)$ such that $\lvert\delta_{G.F}(v)\rvert \leq 1, \forall\, v \in V(G)$ |
| *perfect matching* | $F \subset E(G)$ such that $\lvert\delta_{G.F}(v)\rvert = 1, \forall\, v \in V(G)$ |
| *2-factor* | $F \subset E(G)$ such that $\lvert\delta_{G.F}(v)\rvert = 2, \forall\, v \in V(G)$ |
| *vertex cover* | set of vertices meeting all edges |
| *vertex packing* | set of vertices that induce a subgraph with no edges |
| *Hamiltonian tour* | cycle meeting all vertices of $G$ |

## A.4 Digraphs

(In what follows, $G$ is a finite directed graph.)

| Notation / Term | Definition |
|---|---|
| $V(G)$ | the vertex set of $G$ |
| $E(G)$ | the edge set of $G$ |
| $t(e)$ | the tail of $e \in E(G)$ |
| $h(e)$ | the head of $e \in E(G)$ |
| $\delta_G^+(v)$ | the edges with tail $v$ and head not $v$ $[v \in V(G)]$ |
| $\delta_G^-(v)$ | the edges with head $v$ and tail not $v$ $[v \in V(G)]$ |
| $\delta_G^+(S)$ | the edges with tail in $S$ and head not in $S$ $[S \subset V(G)]$ |
| $\delta_G^-(S)$ | the edges with head in $S$ and tail not in $S$ $[S \subset V(G)]$ |
| *strict digraph* | no loops or identical edges |
| $A(G)$ | $0/\pm 1$-valued vertex-edge incidence matrix of $G$ |
| *dicycle* | directed cycle |
| *directed Hamiltonian tour* | dicycle meeting all vertices of $G$ |

# References

## Background Reading


- D. Bertsimas and J.N. Tsitsiklis (1997), *Introduction to Linear Optimization*, Athena Scientific.
- V. Chvátal (1983), *Linear Programming*, Freeman.
- M. Garey and D. Johnson (1979), *Computers and Intractability: A Guide to the Theory of NP-Completeness*, Freeman.
- D. West (1996), *Introduction to Graph Theory*, Prentice-Hall.


## Further Reading

Of particular note is the magnificient three volume set by Schrijver (2003) which is an encylopedic treatment of combinatorial optimization. Also, the collection edited by Graham, Grötschel and Lovász (1995) is a valuable reference.


- A. Aarts and J.K. Lenstra, eds. (1997), *Local Search in Combinatorial Optimization*, Wiley.
- R.K. Ahuja, T.L. Magnanti, and J.B. Orlin (1993), *Network Flows*, Prentice-Hall.
- K. Anstreicher, N. Brixius, J.-P. Goux, and J. Linderoth (2002), "Solving large quadratic assignment problems on computational grids," *Mathematical Programming* **91**, 563–588.
- D. Applegate, R. Bixby, V. Chvátal, and W. Cook (1998), "On the solution of traveling salesman problems," in *Proceedings of the International Congress of Mathematicians, Vol. III* (Berlin, 1998). Documenta Mathematica 1998, Extra Vol. III, 645–656 (electronic).
- A. Björner, M. Las Vergnas, B. Sturmfels, N. White, and G. Ziegler (1999), *Oriented Matroids*, second edition, Cambridge University Press.
- W. Cook and P. Seymour (2003), "Tour merging via branch-decomposition," *INFORMS Journal on Computing* **15**, 233–248.
- G. Cornuéjols (2001), *Combinatorial Optimization: Packing and Covering*, Society for Industrial and Applied Mathematics.
- E.V. Denardo (1982), *Dynamic Programming: Models and Applications*, Prentice-Hall.
- R.L. Graham, M. Grötschel, and L. Lovász, eds. (1995), *Handbook of Combinatorics*, Elsevier.
- M. Grötschel, L. Lovász, and A. Schrijver (1988), *Geometric Algorithms and Combinatorial Optimization*, Springer-Verlag.

- S. Iwata, L. Fleischer, and S. Fujishige (1999), "A strongly polynomial-time algorithm for minimizing submodular functions," *Sūrikaisekikenkyūsho Kōkyūroku* **1120**, 11–23.
- L. Ladányi, T.K. Ralphs, and L.E. Trotter, Jr. (2001), "Branch, cut, and price: Sequential and parallel," in M. Jünger and D. Naddef, eds., *Computational Combinatorial Optimization: Optimal or Provably Near-Optimal Solutions, Lecture Notes in Computer Science* **2241**, 223–260.
- E.L. Lawler, J.K. Lenstra, A.H.G. Rinnooy Kan, and D.B. Shmoys, eds. (1985), *The Traveling Salesman Problem : A Guided Tour of Combinatorial Optimization*, Wiley.
- J. Lee (2001), "Maximum entropy sampling," in A.H. El-Shaarawi and W.W. Piegorsch, eds., *Encyclopedia of Environmetrics*, Wiley.
- J. Lee and J. Ryan (1992), "Matroid applications and algorithms," *ORSA (now, INFORMS) Journal on Computing* **4**, 70–98.
- H.W. Lenstra, Jr. (1983), "Integer programming with a fixed number of variables," *Mathematics of Operations Research* **8**, 538–548.
- J.T. Linderoth and M.W.P. Savelsbergh (1999), "A computational study of branch and bound search strategies for mixed integer programming," *INFORMS Journal on Computing* **11**, 173–187.
- L. Lovász and M.D. Plummer (1986), *Matching Theory*, Akademiai Kiado.
- F. Margot (2003), "Exploiting orbits in symmetric ILP," *Mathematical Programming* **98**, 3–21.
- A. Martin (2001), "General mixed integer programming: Computational issues for branch-and-cut algorithms," in M. Jünger and D. Naddef, eds., *Computational Combinatorial Optimization: Optimal or Provably Near-Optimal Solutions, Lecture Notes in Computer Science* **2241**, 1–25.
- S.T. McCormick (2003), "Submodular function minimization," in K. Aardal, G. Nemhauser, and R. Weismantel, eds., to appear in the *Handbook on Discrete Optimization*, Elsevier.
- G.L. Nemhauser and L.A. Wolsey (1988), *Integer and Combinatorial Optimization*, Wiley.
- J. Oxley (1992), *Matroid Theory*, Oxford University Press.
- A. Recski (1988), *Matroid Theory and its Applications*, Springer-Verlag.
- A. Schrijver (1986), *Theory of Linear and Integer Programming*, Wiley.
- A. Schrijver (2000), "A combinatorial algorithm minimizing submodular functions in strongly polynomial time," *Journal of Combinatorial Theory, Series B* **80**, 346–355.
- **!** A. Schrijver (2003), *Combinatorial Optimization: Polyhedra and Efficiency*, Springer-Verlag.
- M. Todd (2002), "The many facets of linear programming," *Mathematical Programming* **91**, 417–436.
- V.V. Vazirani (2001), *Approximation Algorithms*, Springer-Verlag.
- W. Whiteley (1992), "Matroids and rigid structures," in N. White, ed., *Matroid Applications*, Encyclopedia of Mathematics and Its Applications, Volume 40, 1–53, Cambridge University Press.
- G.M. Ziegler (1994), *Lectures on Polytopes*, Springer-Verlag.

# Indexes

## Examples



## Exercises



## Problems

**Results**

**Algorithms**